MULTIFUNCTIONAL
MATERIALS AND MODELING

MULTIFUNCTIONAL MATERIALS AND MODELING

Edited by

Mikhail A. Korepanov, DSc, and Alexey M. Lipanov, DSc

Series Editors-in-Chief:
Vjacheslav B. Dement'ev, DSc, and Vladimir I. Kodolov, DSc

Gennady E. Zaikov, DSc, and A. K. Haghi, PhD
Reviewers and Advisory Board Members

Apple Academic Press Inc. | Apple Academic Press Inc.
3333 Mistwell Crescent | 9 Spinnaker Way
Oakville, ON L6L 0A2 | Waretown, NJ 08758
Canada | USA

© 2016 by Apple Academic Press, Inc.

First issued in paperback 2021

Exclusive worldwide distribution by CRC Press, a member of Taylor & Francis Group

No claim to original U.S. Government works

ISBN-13: 978-1-77463-229-1 (pbk)
ISBN-13: 978-1-77188-087-9 (hbk)

Library and Archives Canada Cataloguing in Publication

Multifunctional materials and modeling / edited by Mikhail. A. Korepanov, DSc, and Alexey M. Lipanov, DSc; Gennady E. Zaikov, DSc, and A.K. Haghi, PhD, reviewers and advisory board members.

(Innovations in chemical physics and mesoscopy)
Includes bibliographical references and index.
ISBN 978-1-77188-087-9 (bound)
1. Materials--Research. 2. Materials--Thermal properties--Research. 3. Nanostructured materi-als-- Research. 4. Surface chemistry. I. Lipanov, A. M. (Alekseï Matveevich), author, editor II. Korepanov, Mikhail. A., author, editor III. Series: Innovations in chemical physics and mesoscopy

TA404.2.M84 2015   620.1'1   C2015-903187-7

Library of Congress Cataloging-in-Publication Data

Multifunctional materials and modeling / [edited by] Mikhail. A. Korepanov, DSc, and Alexey M. Lipanov, DSc.

pages cm. -- (Innovations in chemical physics and mesoscopy)
Includes bibliographical references and index.
ISBN 978-1-77188-087-9 (alk. paper)
1. Materials--Research. 2. Materials--Thermal properties--Research. 3. Nanostructured materials--Research. 4. Surface chemistry. I. Korepanov, Mikhail. A. II. Lipanov, A. M. (Aleksei Matveevich)

TA404.2.M85 2015   541'.2--dc23   2015016729

Apple Academic Press also publishes its books in a variety of electronic formats. Some content that appears in print may not be available in electronic format. For information about Apple Academic Press products, visit our website at **www.appleacademicpress.com** and the CRC Press website at **www.crcpress.com**

ABOUT THE SERIES
INNOVATIONS IN CHEMICAL
PHYSICS AND MESOSCOPY

The Innovations in Chemical Physics and Mesoscopy book series publishes books containing original papers and reviews as well as monographs. These books and monographs will report on research developments in the following fields: nano-chemistry, mesoscopic physics, computer modeling, and technical engineering, including chemical engineering. The original papers, reviews, and monographs are submitted by the authors on scientific trends reported in the Russian journal *Chemical Physics & Mesoscopy* as well as on related trends of analogous international magazines. The books in this series will prove very useful for academic institutes and industrial sectors interested in advanced research.

Editors-in-Chief:
Vjacheslav B. Dement'ev, DSc
Professor and Director, Institute of Mechanics, Ural Division, Russian Academy of Sciences
e-mail: demen@udman.ru

Vladimir Ivanovitch Kodolov, DSc
Professor and Head of Chemistry and Chemical Technology Department at M. T. Kalashnikov Izhevsk State Technical University; Director of BRHE center of Chemical Physics and Mesoscopy, Udmurt Scientific Centre, Russian Academy of Sciences
e-mail: kodol@istu.ru; vkodol.av@mail.ru

Editorial Board:
A. K. Haghi, PhD
Professor, Associate Member of University of Ottawa, Canada;
Member of Canadian Research and Development Center of Science and Culture
e-mail: akhaghi@yahoo.com

Victor Manuel de Matos Lobo, PhD
Professor, Coimbra University, Coimbra, Portugal

Richard A. Pethrick, PhD, DSc
Research Professor and Professor Emeritus, Department of Pure and Applied Chemistry, University of Strathclyde, Glasgow, UK

Eli M. Pearce, PhD
Former President, American Chemical Society; Former Dean, Faculty of Science and Art, Brooklyn Polytechnic University, New York, USA

Mikhail A. Korepanov, DSc
Research Senior of Institute of Mechanics, Ural Division, Russian Academy of Sciences

Alexey M. Lipanov, DSc
Professor and Head, Udmurt Scientific Center, Russian Academy of Sciences;

Editor-in-Chief, *Chemical Physics & Mesoscopy* (journal)

Gennady E. Zaikov, DSc
Professor and Head of the Polymer Division at the N. M. Emanuel Institute of Bio-chemical Physics, Russian Academy of Sciences

BOOKS IN THE SERIES

Multifunctional Materials and Modeling
Editors: Mikhail A. Korepanov, DSc, and Alexey M. Lipanov, DSc
Reviewers and Advisory Board Members: Gennady E. Zaikov, DSc, and
A. K. Haghi, PhD

Mathematical Modeling and Numerical Methods in Chemical Physics and Mechanics
Ali V. Aliev, DSc, Olga V. Mishchenkova, PhD
Editor: Alexey M. Lipanov, DSc

ABOUT THE EDITORS

Mikhail A. Korepanov, DSc

M. A. Korepanov, DSc, is Professor at Udmurt State University, Izhevsk, Russia, where he teaches thermodynamics, theory of combustion, and physical-chemical fluid mechanics. He is editor-in-chief of the journal *Chemical Physics and Mesoscopy* (Russia) and has been named an Honorable Scientist of the Udmurt Republic. His research interests are hydrodynamics, chemical thermodynamics, and numerical simulation of processes in technical systems and natural phenomena.

Alexey M. Lipanov, DSc

A. M. Lipanov, DSc, is Professor at the M. T. Kalashnikov Izhevsk State Technical Institute in Izhevsk, Russia, where he teaches on applied fluid mechanics, internal ballistics of rocket engines, and numerical simulation of processes in technical systems. He is senior editor of the journal *Chemical Physics and Mesoscopy* (Russia) and is on the editorial boards of several international journals. He has published over 500 articles and 12 books and holds 65 patents. He has received several awards, including Honorable Scientist of the Russian Federation and Honorable Worker of High Education of the Russian Federation. His research interests include hydrodynamics, turbulence, computational mathematics, intrachamber processes in solid propellant rocket engines, and numerical simulation of processes in technical systems and natural phenomena.

Vjacheslav B. Dement'ev, DSc

Vjacheslav B. Dement'ev, DSc, is a Professor . T. Kalashnikov Izhevsk State Technical University in Izhevsk, Russia He was formerly Director of the Institute of Mechanics, Ural Division, at the Russian Academy of Sciences. The author of over 200 articles, 70 patents, and six books, he was honored with the title Honorable Scientist of Udmurt Republic, Order of Merit. His particular research interests include Processes Chemical Physics For the Production of Metallic Substances and Materials, Nanotechnology and Engineering in Nanomaterials.

Vladimir I. Kodolov, DSc

Vladimir I. Kodolov, DSc, is Professor and Head of the Department of Chemistry and Chemical Technology at M. T. Kalashnikov Izhevsk State Technical University in Izhevsk, Russia, as well as Chief of Basic Research at the High Educational Center of Chemical Physics and Mesoscopy at the Udmurt Scientific Center, Ural

Division at the Russian Academy of Sciences. He is also the Scientific Head of the Innovation Center at the Izhevsk Electromechanical Plant in Izhevsk, Russia. He is Vice Editor-in Chief of the Russian *journal Chemical Physics and Mesoscopy* and also is a member of the editorial boards of several Russian journals. He has been distinguished as Honored Scientist of the Udmurt Republic, Honored Scientific Worker of the Russian Federation, Honorary Worker of Russian Education, and also Honorable Academician of the International Academic Society. He has over 800 publications to his name, including monographs, articles, reports, reviews, and patents.

REVIEWERS AND ADVISORY BOARD MEMBERS

A. K. Haghi, PhD

A. K. Haghi, PhD, holds a BSc in urban and environmental engineering from University of North Carolina (USA); a MSc in mechanical engineering from North Carolina A&T State University (USA); a DEA in applied mechanics, acoustics and materials from Université de Technologie de Compiègne (France); and a PhD in engineering sciences from Université de Franche-Comté (France). He is the author and editor of 65 books as well as 1000 published papers in various journals and conference proceedings. Dr. Haghi has received several grants, consulted for a number of major corporations, and is a frequent speaker to national and international audiences. Since 1983, he served as a professor at several universities. He is currently Editor-in-Chief of the *International Journal of Chemoinformatics and Chemical Engineering* and *Polymers Research Journal* and on the editorial boards of many international journals. He is a member of the Canadian Research and Development Center of Sciences and Cultures (CRDCSC), Montreal, Quebec, Canada.

Gennady E. Zaikov, DSc

Gennady E. Zaikov, DSc, is Head of the Polymer Division at the N. M. Emanuel Institute of Biochemical Physics, Russian Academy of Sciences, Moscow, Russia, and Professor at Moscow State Academy of Fine Chemical Technology, Russia, as well as Professor at Kazan National Research Technological University, Kazan, Russia. He is also a prolific author, researcher, and lecturer. He has received several awards for his work, including the Russian Federation Scholarship for Outstanding Scientists. He has been a member of many professional organizations and on the editorial boards of many international science journals.

CONTENTS

PART I: COMPUTATIONAL MODELING

LIST OF CONTRIBUTORS

V. V. Aksenova
Physical-Technical Institute, Ural Branch, Russian Academy of Science, Izhevsk

D. V. Anisimov
Udmurt State University, Izhevsk

N. V. Baranovskiy
National Research Tomsk Polytechnic University, Tomsk

Ya V. Bayankin
Physical-Technical Institute, Ural Branch, Russian Academy of Science, Izhevsk

Ya. B. Benderskiy
Kalashnikov Izhevsk State Technical University, Izhevsk

D. A. Bezbabnyi
Far Eastern Federal University, Vladivostok

A. A. Bolkisev
Institute of Mechanics Ural Branch, Russian Academy of Science, Izhevsk

N. S. Buldakova
Udmurt State University, Izhevsk

P. V. Bykov
Physical-Technical Institute, Ural Branch, Russian Academy of Science, Izhevsk

S. G. Bystrov
Physical-Technical Institute, Ural Branch, Russian Academy of Science, Izhevsk

I. M. Chernev
Institute of Automation and Control Processes, Far East Branch, Russian Academy of Science, Vladivostok

T. M. Chmereva
Orenburg State University, Orenburg

A. A. Demchenko
Ufa State Petroleum Technological University, Ufa

M. V. Demchenko
Ufa State Petroleum Technological University, Ufa

V. B. Dementyev
Institute of Mechanics, Ural Branch, Russian Academy of Science, Izhevsk

A. D. Dmitriev
Orenburg State University, Orenburg

G. A. Dorofeev
Physical-Technical Institute, Ural Branch Russian Academy of Science, Izhevsk

S. A. Dotsenko
Institute of Automation and Control Processes, Far East Branch, Russian Academy of Science, Vladivostok

K. N. Galkin
Institute of Automation and Control Processes, Far East Branch, Russian Academy of Science, Vladivostok

N. G. Galkin
Institute of Automation and Control Processes, Far East Branch, Russian Academy of Science, Vladivostok

F. Z. Gilmutdinov
Physical-Technical Institute, Ural Branch, Russian Academy of Science, Izhevsk

D. O. Glushkov
National Research Tomsk Polytechnic University, Tomsk

Yu N. Isupov
Udmurt State University, Izhevsk

G. A. Ivanov
Kemerovo State University, Kemerovo

A. I. Kalugin
Udmurt State University, Izhevsk

O. M. Kanunnikova
Physical-Technical Institute, Ural Branch, Russian Academy of Science, Izhevsk

A. V. Khaneft
Kemerovo State University, Kemerovo

E. V. Kharanzhesky
Udmurt State University, Izhevsk

A. V. Kholzakov
Physical-Technical Institute, Ural Branch, Russian Academy of Science, Izhevsk

A. Y. Kirsanov
Ural Federal University named after the first President of Russia B. N. Yeltsin, Ekaterinburg

I. N. Klimova
Physical-Technical Institute, Ural Branch, Russian Academy of Science, Izhevsk

V. F. Kobziev
Udmurt State University, Izhevsk

V. I. Kodolov
Kalashnikov Izhevsk State Technical University, Basic Research High Educational Centre at Udmurt Scientific Centre of Ural Division, Russian Academy of Sciences, Izhevsk

A. A. Kolotov
Physical-Technical Institute, Ural Branch, Russian Academy of Science, Izhevsk

K. A. Kopylov
Kalashnikov Izhevsk State Technical University, Izhevsk

G. A. Korablev
Basic Research and Educational Center, Udmurt Scientific Center, Ural Branch Russian Academy of Science; Izhevsk State Agricultural Academy, Izhevsk

M. A. Korepanov
Institute of Mechanics, Ural Branch, Russian Academy of Science, Izhevsk

V. I. Kornev
Udmurt State University, Izhevsk

V. I. Kozhevnikov
Institute of Mechanics, Ural Branch, Russian Academy of Science, Izhevsk

G. V. Kozlov
Kh. M. Berbekov Kabardino-Balkarian State University, Nalchik

M. G. Kucherenko
Orenburg State University, Orenburg

V. I. Ladyanov
Physical-Technical Institute, Ural Branch, Russian Academy of Science, Izhevsk

A. M. Lipanov
Institute of Mechanics Ural Branch, Russian Academy of Science, Izhevsk

N. V. Lomova
Udmurt State University, Izhevsk

A. N. Lubnin
Physical-Technical Institute, Ural Branch Russian Academy of Science, Izhevsk

S. S. Makarov
Institute of Mechanics, Ural Branch, Russian Academy of Science, Izhevsk

T. M. Makhneva
Institute of Mechanics, Ural Branch, Russian Academy of Science, Izhevsk

V. F. Markov
Ural Federal University named after the first President of Russia B.N.Yeltsin, Ekaterinburg

L. N. Maskaeva
Ural Federal University named after the first President of Russia B.N.Yeltsin, Ekaterinburg

A. V. Markidonov
Kuzbass State Technical University, Novokuznetsk branch, Novokuznetsk

D. A. Merzlyakov
Udmurt State University, Izhevsk

S. S. Mikhailova
Physical-Technical Institute, Ural Branch, Russian Academy of Science, Izhevsk

A. K. Mikitaev
Kh.M. Berbekov Kabardino-Balkarian State University, Nalchik

V. V. Muhgalin
Physical-Technical Institute, Ural Branch, Russian Academy of Science, Izhevsk

R. R. Mulyukov
Institute for Metals Super plasticity Problems, Russian Academy of Science, Ufa

O. S. Nabokova
Institute of Mechanics, Ural Branch, Russian Academy of Science, Izhevsk

A. V. Obukhov
Udmurt State University, Izhevsk

E. P. Pavlovskaya
Kuzbass State Technical University, Novokuznetsk branch, Novokuznetsk

V. G. Petrov
Institute of Mechanics, Ural Branch, Russian Academy of Science, Izhevsk

N. I. Petrova
Udmurt State University, Izhevsk

Ya. A. Polyotov
Kalashnikov Izhevsk State Technical University, Izhevsk

A. G. Ponomarev
Physical-Technical Institute, Ural Branch, Russian Academy of Science, Izhevsk

G. V. Sapozhnikov
Udmurt State University, Izhevsk

E. V. Safonov
South Ural State University, Chelyabinsk

I. N. Shabanova
Physical-Technical Institute, Ural Branch, Russian Academy of Science, Izhevsk

A. V. Shalimova
Institute for Metals Super plasticity Problems, Russian Academy of Science, Ufa

A. V. Severyukhin
Institute of Mechanics Ural Division, Russian Academy of Science, Izhevsk

A. V. Sisanbaev
Institute for Metals Super plasticity Problems, Russian Academy of Science, Ufa

A. L. Slonov
Kh. M. Berbekov Kabardino-Balkarian State University, Nalchik

V. V. Sobolev
Udmurt State University, Izhevsk

Val. V. Sobolev
Kalashnikov Izhevsk State Technical University, Izhevsk

M. D. Starostenkov
I. I. Polzunov Altai State Technical University, Barnaul

P. A. Strizhak
National Research Tomsk Polytechnic University, Tomsk

D. V. Strugova
Orenburg State University, Orenburg

M. A. Shumilova
Institute of Mechanics, Ural Branch, Russian Academy of Science, Izhevsk

S. V. Suvorov
Institute of Mechanics Ural Division, Russian Academy of Science, Izhevsk

N. S. Terebova
Physical-Technical Institute, Ural Division, Russian Academy of Science, Izhevsk

V. A. Trapeznikov
Physical-Technical Institute, Ural Division, Russian Academy of Science, Izhevsk

V. V. Trineeva
Institute of Mechanics, Ural Division, Russian Academy of Science, Izhevsk

A. V. Vakhrushev
Institute of Mechanics Ural Division, Russian Academy of Science

V. L. Vorobiev
Physical-Technical Institute, Ural Division, Russian Academy of Science, Izhevsk

G. E. Zaikov
N.M. Emanuel Institute of Biochemical Physics, Russian Academy of Science, Moskow

A. V. Zakharevich
National Research Tomsk Polytechnic University, Tomsk

A. V. Zhikharev
Physical-Technical Institute, Ural Division, Russian Academy of Science, Izhevsk

L. R. Zubairov
Institute for Metals Super plasticity Problems, Russian Academy of Science, Ufa

LIST OF ABBREVIATIONS

AES	Auger Electron Spectroscopy
AFM	Atomic Force Microscopy
BC	Boundary Conditions
BZ	Brillouin Zone
CS	Condensed Substances
CSP	Condensed Solid Propellants
DFT	Density Functional Theory
DHS	Double Hetero Structures
EELS	Electron Energy Loss Spectroscopy
eV	Electron-Volts
EVA	Ethylene-Vinyl Acetate Copolymer
FDS	Fine-Dispersed Suspension
FP LMTO	Full-Potential Method of Linear Muffin-Tin Orbitals
FS	Fine Suspensions
FSM	Fourier Spectrometer
GGA	Generalized Gradient Approximations
H/C	Hydrogen-Carbon
HTPE	Hydroxy Terminated Polyether
HTTT	High-Temperature Thermo mechanical Treatment
ICDD	International Centre for Diffraction Data
IR	Infrared Spectroscopy
IRF	Refractive Index Measurement
LDA	Local Electron Density
LHC	Liquid Hydrocarbons
MBE	Molecular Beam Epitaxy
MM	Mechanical Milling
MO LCAO	Molecular Orbital Linear Combination of Atomic Orbitals
NC	Nanocomposite
PEPA	polyethylene Polyamine
PMMA	Polymethylmethacrylate
PVA	Polyvinyl Alcohol
RDE	Reactive Deposition Epitaxy
SF	Stacking Faults
SIC	Self-Interaction Correction
SOI	Spin-Orbit Interaction
SPE	Solid Phase Epitaxy

TEM	Transmission Electron Microscopy
THP	Torsion under High Pressure
TiC	Titanium Carbide
UVB	Upper Valence Band
XPEM	X-ray Photoelectron Microscopy
XPS	X-ray Photoelectron Spectroscopy

LIST OF SYMBOLS

A	specific surface of pores
a	nanoreactor activity
A_{ef}	effective atomic mass
b_1 and b_2	empirical constants
b	thickness of the near-surface layer
C	Stefan-Boltzmann constant
$C_xH_{y(liq)}$	hydrocarbons, toluene and n-heptane
D	coefficient of radial diffusion of oxygen molecules
d	dimension of the Euclidean space
d	pore diameter
D_i^M	effective diffusion coefficient of i-th species in the mixture
E	thermal decomposition activation energy
E_k	electron kinetic energy
F	Faraday number
h	film thickness
H_i	enthalpy of i-th species
H_f	specific heat of fusion
i	number of the atom
I(t)	energy flux density of the electron beam
j(0, t)	current density of electron at the input into the solid
k	value corresponding to specific process rate
k_B	Boltzmann constant
k(x)	attenuation coefficient
m	nanostructure mass
m_i	mass of the atom
m_r	reduced mass
N	total number of atoms in the system
N_p	number of moles of the product produced in nanoreactor
N_r	number of moles of reagents atoms (ions) participating in the process which filled the nanoreactor
n_i	amount of i-th species
$n_T(t)$	surface concentration of t centers
P_0	parameter of atom external valence electrons
P_{CD}	parameter of countdown
P_i-P_0	parameters of each element in the system

p	pressure
Q	thermal effect of reaction per one mass unit of substance
$\dot{Q}_{chem}$	heat effect of chemical reactions
$\dot{Q}_{heat}$	absorbed heat flux
$\dot{Q}_0$	surface heat flux
q_f	power density of the laser irradiation, W/m²
R	dimensional characteristic of tom or chemical bond
Re	Reynolds number
rf	radius of the focal spot, m
rk	covalence radii
Ref	effective track length of electrons
Rre	coefficient of the reflectivity of the surface of the material
Rf	specific flow resistance force
$\vec{r}_i$	radius vector atom
S	surface of nanoreactor walls
T	temperature
Tf	fusion temperature
$T0$	initial temperature
Tm	temperature of "nose" of the curve of ferrite or pearlite
TS	upper temperature limit of transformation
U	potential energy of the system
U_b	potential changes in bond length
U_{ej}	potential changes planar groups
U_{qq}	potential electrostatic interaction
U_{hb}	potential hydrogen bonding
U_{VW}	potential of the van der Waals interaction
$U_0(t)$	accelerating stress of electron beam generator
V	volume in which chemical reactions take place
$\tilde{v}$	effective flow rate
V1, V2	water volume passed through the soil samples with different initial concentration of pollutant
W	share of nanoproduct obtained in nanoreactor
w	volume density of absorbed energy
Z	reaction frequency factor
z	number of electrons participating in the process
Zef	effective atomic number
Z*	nucleus effective charge

GREEK SYMBOLS

α	heat transfer coefficient
$\alpha\Gamma K$	coefficient whose value 1.5
$\Delta U1$	potential energies of material points on elementary region of interactions
ΔU	mutual potential energy of the interactions
ΔP	difference between the P0-parameter of i orbital
$\Delta\varphi$	difference of potentials at the boundary "nano reactor wall reactive mixture
ε	emissivity of the surface
$\varepsilon 0Sd$	multiplication of surface layer energy by its thickness
$\varepsilon 0V$	energy of nanoreactor volume unit
εS	surface energy reflecting the energy of interaction of reagents with nano reactor walls
εV	nanoreactor volume energy
$g_\Delta(\vartheta, t)$	distribution function of singlet-excited oxygen molecules relative to system of T centers in the pore
$\Lambda(x)$	distribution of absorbed energy density along the crystal
η	conversion level of the explosive
τ	duration
τm	incubation period of the decay at temperature Tm
$\theta(\vartheta - \vartheta_0)$	Heaviside theta-function
$\dot{m}$	mass inflow rate of gases due to binder decomposition
$T\Pi OB$	body surface temperature
$TOKP$	ambient temperature
φ_f^3	dimensionless volume share of porous CS particles
φ_0^3	dimensionless volume share of oxidant (air)
φ_f^1	dimensionless volume of combustible gases
φ_0^1	dimensionless volume of oxidant (air)
ρ	thermal conductivity coefficient
$\aleph$	thermal conductivity of the material, $W/(m\times K)$
$\gamma_{ef} = 1,2$	ration of specific heat capacity of the solid and plasma
λ	eigenvalue of Fokker-Planck operator
ϕ	porosity
v	gas velocity
υ	crystallinity degree
U_θ	potential changes in the angle between the bonds

U_φ	potential changes in the torsion angle connection
nкс	wave numbers of chemical bonds changing during the process
nнс	corresponds to wave numbers of initial state of chemical bonds
vsc	scanning speed
υк	velocity of nanostructure oscillations
РЕ,λ	spatial-energy parameter of quantum transition
ω_i	mass fraction of i-th species
χ	affinity for electron

PREFACE

This book is the first volume from the new series Innovations in Chemical Physics and Mesoscopy. This volume includes some important work from the Institute of Mechanics at Ural Division of Russian Academy of Sciences.

This important book presents a valuable collection of new research and new trends in nanomaterials, mesoscopy, quantum chemistry, and chemical physics processes, including these topics:

1. combustion and explosion processes
2. mathematical modeling of physical-chemical processes
3. clusters, nanostructures and nanostructured materials
4. interfaced layers and interaction processes within these layers
5. quantum-chemical calculations
6. nonlinear kinetic phenomena
7. the devices and equipment of nanoelectronics

The chapters are divided into three major sections: computational modeling, surface and interface investigations, and nanochemistry, nanomaterials, and nanostructured materials and present some of the most important information on these trends. The book presents 28 papers from experts in the field, and also provides a valuable review on results and new data from investigations concerning metal/carbon nanocomposites.

The book will be of value and interest to researchers, professors, postgraduate students, and industry professionals.

We thank the Institute of Mechanics at Ural Division of Russian Academy of Sciences as well as the individual chapter authors for their valuable research and work in the field.

CHAPTER 1

THE CALCULATIONS OF THE COMPLEX OF THE RED MERCURIC IODIDE FUNDAMENTAL OPTICAL FUNCTIONS

D. V. ANISIMOV, V. V. SOBOLEV, and V. VAL. SOBOLEV

CONTENTS

1.1 INTRODUCTION

The tetragonal di iodide mercuric HgI_2 is crystallized in the lattice with symmetry D_{4h}^{15} –P4$_2$/nmc [1]. It is known as the most perspective material for the fabricating high-resolution x-ray and γ-ray detectors [2, 3]. But his electronic structure and optical properties are investigated relatively small [4–7].

The reflectivity spectra of the monocrystals were measured at 4.2 K for the polarizations E⊥c, E∥c in the range 2.25 to 2.65 eV, E⊥c in the range 538 to 526 nm and 1.8 to 6.0 eV [8], in the range 1.9 to 10 eV at 100 K for E⊥c, 2 to 6 eV at 15 K and E⊥c, E∥c [9]. The samples are easily cleaved perpendicular to the optical c-axis. But the non-perfect surface if the as-grown is have this axis. Therefore, the measurements were carry out for the E⊥c the cleavage and for the E∥c – on the as-grown samples. The long-wavelength intensive excitonic reflectivity band is strongly polarized. Therefore, it is visible only at the E⊥c. More over two band groups are in the bigger energy 2 to 6 and 6 to 10 eV, which separated by the intensive minimum and consisted from the many highly overlapped partially polarized maxima.

Theoretical bands and spectra of $e_2(E)$ of the tetragonal HgI_2 are calculated in [2, 10, 11]. The upper valence band with the width smaller than 4 eV consists from the five-duplet bands. Lower on the ~1 to 3 eV is the separate duplet and still lower on the ~6.0 to 7.5 eV and ~12 eV are two complex Hgd- and Js-bands correspondingly. The lower conduction band is the duplet and separated and above it is many strongly disperse and overlapped bands. Therefore, it is naturally that the theoretical spectra of $e_2(E)$ and R(E) are have very many peaks the concrete nature of which is not considerate.

The experimental investigations of the electronic structure of the crystal and the analysis of the theoretical calculations are needed in the extensive complex of the comprehensive measurements in the wide energy region, including not only the optical spectra but also the photo emissive spectra of the perfect and oriented monocrystals. Traditionally it is mainly may be only for the reflectivity spectra. Aber such measurements are contained raw information with energy of the reflectivity maxima which cliffs from the energy of the band transitions very much.

The purpose of the communication is in the obtaining of the new information about the optical properties and electronic structure of the HgI_2 crystals in the wide energy region: to assess the spectra of the common complex of the fundamental optical functions, decomposition of the e_2 and $-\mathrm{Im}\ \varepsilon^{-1}$ into the elemental components with their main parameters and the determination of the main peculiarities of the optical functions and their components.

1.2 CALCULATION METHODS

It is generally accepted that the most complete information of the optical properties is contents in the 15 fundamental optical functions [12, 13]: reflectivity (R) and absorption (α) coefficients; index of refractive (n) and absorption (k); imaginary (e_2) and real (e_1) parts of the dielectric permittivity ε; real ($\text{Re}\varepsilon^{-1}$, $\text{Re}(1+\varepsilon)^{-1}$) and imaginary parts ($-\text{Im } \varepsilon^{-1}$, $-\text{Im}(1+\varepsilon)^{-1}$) of the inverse dielectric functions ε^{-1} and $(1+\varepsilon)^{-1}$; integral function of the combined density of states I_c, which is equal to $e_2 E^2$ at the constant intensity of the transitions with the universal factors; effective quantity of the valence electrons $n_{eff}(E)$, participated in transitions to energy E, which calculated by the four methods with the known spectra of e_2, k, $-\text{Im } \varepsilon^{-1}$, $-\text{Im}(1+\varepsilon)^{-1}$; effective dielectric permittivity e_{eff} and other.

In the wide energy region usually only experimental reflectivity spectrum is know. Therefore, the spectra of the last functions were calculated on the basis of R(E) using the computer program which applied the Kramers-Kronig integral correlation and analytical relations between optical functions.

The complex of the optical functions usually is calculated using the known reflectivity spectra R(E) in the wide energy range. Furth more is obtained the decomposition of the e_2 and $-\text{Im } \varepsilon^{-1}$ spectra into the components and detecting the main parameters of the components (the energy of the maximum E_i and semi width H_i, band area S_i amplitude I_i and oscillators strength) by the method of the combined diagram Argand. Applied calculational methods are published in Refs. [12–17].

1.3 RESULT AND DISCUSSION

The reflectivity spectra of the cleaved HgI_2 were measured in the range 1.8 to 10 eV at 100 K [9] and 1.8 to 6 eV at 4.2 K [8] for E⊥c polarization. The cleaved samples with optical axis in the surface, were very incomplete. Therefore, their reflectivity spectra R(E) for E⊥c at 15 K were highly underestimated un-uniformly on the energy in the 1.8 to 6 eV. The R(E) spectrum for E‖c was may be highly deformed without application for the calculations of other optical functions. Therefore, in the communication we used the spectra R(E) at 100 K in the energy range 1.8 to 10 eV [9] and at 4.2 K from 2.33 to 2.36 eV [8]. The extrapolation R(E) in the energy $E < $ 1.8 eV is using nearly constant and in the $E > 10$ eV is by the Philipp-Taft method [12, 13] for $R \sim E^{-p}$. The both extrapolations are accomplished at the keeping of the sum rule for the n(E) in the $E > 12$ eV: the n(E) curve must be very nearly $n \approx 1$ and raised with the energy.

Experimental reflectivity spectrum has the intensive and narrow exciton peak (№ 1), 11 maxima and 5 shoulders (Fig. 1.1a, Table 1.1). Their analogs are visible in the calculated spectra of the majority other optical functions with the small energy displations to the smaller or bigger energy (e_2, k, α, $e_2 E^2$, $-\text{Im } \varepsilon^{-1}$).

The spectral distribution of the structural intensity (maxima and shoulders) are highly depends from the nature of the optical function. Usually the exsitonic curve R(E) at ~2.34 eV is highly asymmetric with the short long-wave shoulder at ~2.342 eV. However, it is sharply narrowing for the other functions and nearly symmetrical Gauss-type with shortwave length asymmetry of the $e_2(E)$, k(E), $\alpha(E)$ and $e_2 E^2(E)$

but with sharply its decreasing for the $e_2(E)$. The maxima of the optical functions are displaced relatively the maximum e_2 for the ~0.7 (k, α), ~0.4 meV (R) in the region of the bigger energy and for ~0.6 meV (e_1, n) in the range of the smaller energy with the longitudinal – transverse energy solilting $\Delta E_{lt} = E(-\text{Im } \varepsilon^{-1}) - E(e_2) \approx 5$ meV. All the other structures are much wider. Therefore, their differences in the maximum energy are considerable higher.

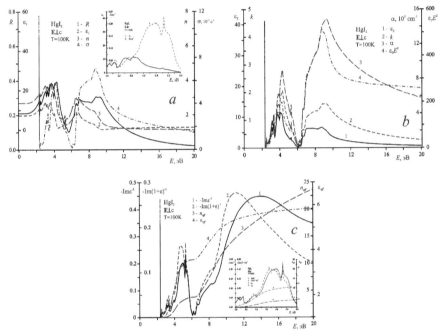

FIGURE 1.1 Experimental spectrum R(E) HgI$_2$ at 100 K for E⊥c (1), calculated spectra e_1 (2), n(3), σ(4) (a), e_2 (1), k(2), α(3), e_2E^2(4) (b), –Im ε^{-1} (1), –Im(1+ε)$^{-1}$ (2), n_{eff}(3), e_{eff}(4) (c); in the onsets are the spectra e_2 (1), –Im ε^{-1} (2) in the 3 to 6 eV (1a), –Im ε^{-1} (1), –Im(1+ε)$^{-1}$ (2), n_{eff} (3), e_{eff}(4) in the 3 to 6 eV (1c).

The spectra of optical functions are consisted from the two groups with the deep minimum at ~5–6 eV. The upper valence band by the theoretical calculations [11] is consisted from the five doublet bands and the doublet lower conduction band is separated from the other conduction bands by the highly interstice. Therefore, it is naturally to suppose that the long wavelength bands group is caused by the transitions from the upper valence band into the lower doublet conduction band and the short long-wave bands group – by the transitions into the higher conduction bands. It is interesting to note that the spectra of long-wave group of the optical functions have five doublet structures with the importantly higher overlapping of the shortwave group than the long-wave group. Therefore, their maxima are visible worse. In the transition from the long-wave group to the shortwave the $R(E)$ and $e_1(E)$, $n(E)$ is reduced into ~1.3 time but the $e_1(E)$ without the negative minimum and is nearly the one at ~9.5 eV. The absorption coefficient with the higher energy is raised to ~8×10^5 and 15×10^5 cm^{-1} in the both bands groups. Such absorption to ~10^6 см$^{-1}$ is characteristically for the highly ionic crystals [3, 4]. In the case of highly reduce of $e_2(E)$ in the energy $E > 10$ eV the $\alpha(E)$ and $e_2E^2(E)$ curves are retained their very high magnitudes.

TABLE 1.1 Energy Maxima and Shoulders (in Bracket) HgI$_2$ Crystal

№	R	ε_1	ε_2	n	k	a	e_2E^2	$-\mathrm{Im}\,\varepsilon^{-1}$	$-\mathrm{Im}(3+\varepsilon)^{-1}$	s
1	2.339	2.338	2.338	2.338	2.393	2.393	2.338	2.344	2.335	2.338
2	2.378	2.368	—	2.368	—	—	—	—	—	—
3	2.49	2.49	2.50	2.49	2.51	2.51	2.51	2.51	2.51	2.50
4	3.09	—3.05	3.13	3.06	3.14	3.14	3.14	3.16	3.16	3.14
5	3.32	3.28	3.33	3.30	3.34	3.35	3.35	3.37	3.36	3.34
6	3.71	3.59	—	3.63	—	—	—	—	—	—
7	(3.8)	—	3.82	(3.8)	3.85	3.87	3.85	—	—	3.84
8	4.12	4.02	4.11	4.04	(4.1)	(4.1)	4.13	3.96	3.96	4.12
9	4.29	—	(4.2)	(4.2)	4.33	4.34	4.27	(4.2)	(4.2)	4.25
10	—	—	—	—	—	—	—	(4.8)	4.76	—
11	—	—	—	—	—	—	—	5.05	(5.0)	—
12	5.29	5.23	5.30	5.23	5.31	5.31	5.30	5.32	5.32	5.30
13	(5.4)	(5.5)	(5.5)	5.47	(5.5)	(5.5)	(5.5)	(5.5)	(5.5)	(5.5)
14	6.18	6.18	6.21	6.17	6.23	6.23	6.24	6.23	6.23	6.23
15	6.59	(6.6)	6.62	6.87	6.64	6.65	6.63	6.72	6.72	6.62
16	6.95	6.88	7.08	—	(7.2)	(7.3)	(7.2)	—	—	(7.1)
17	(8.5)	8.36	—	(8.4)	(8.2)	(8.2)	(8.2)	8.28	8.24	(8.1)
18	8.83	—	8.75	—	8.87	(8.9)	8.79	—	—	8.79
19	—	(9.0)	(9.0)	(9.0)	—	—	(9.0)	(9.0)	(9.0)	(9.0)
20	9.31	—	(9.7)	—	9.31	9.35	—	—	—	—
Eps	—	—	—	—	—	—	—	—	12.2	—
Epv	—	—	—	—	—	—	—	14.2	—	—

The spectra of the characteristic electron energy losses are content very wide and intensive bands with the maxima at E_{pv1} = 14.2 and E_{ps1} = 11 eV caused by the volume and surface plasmons due to the excitation all the collective valence electrons. Such spectra have the one wide and relatively intensive band in the range 4.4 to 5.2 eV with the highly decreasing R(E) and e_2(E) and absence their structures (Fig. 1.1a, c). It is naturally to purpose this band is caused by the excitation of the plasmons due to the collective of the higher valence electrons. The long-wave plasmon band manifestation is distinctively for the all layer crystals [4, 18]. It follows from the calculated spectrum of one n_{eff}(E) (Fig. 1.1c) that the long-wave plasmon band of HgI_2 is caused by the excitation of nearly four valence electrons on the one molecule of the compounds. Also this group of electrons is participated in the formation of the long wavelength band of the optical functions.

Between all structures is visible very narrow peak of the all-optical functions at ~5.3 eV. It may be cause by the transitions of exciton-type due to the same curvature of the lower conduction band and one of valence bands lower than the upper on the ~3 eV in the volume of BZ and is determined by the area of this band, that is by the oscillator field: This unicum method of the direct determination of the oscillators field for the crystal band transition may be applied in the case of very intensive narrow long-wave length exciton line which is not overlapped with the absorption continuum or other exciton lines [19, 20]. With energy the curve n_{eff}(E) for the E > 6.5 eV is increased to ~23 at 20 eV (Fig. 1.1c). But in the one formula unity HgI_2 is 16 valence electrons. In accordance with the theoretical calculations s – and d – bands of Hg are lower the maximum of UVB on the ~5 and 7 eV correspondingly. Partial electrons of these bands are take in formation of n_{eff}(E).

The optical functions are caused the different effects, for example, e_2 – by the dissipation of the energy of the wave, k – by the damping of the amplitude, –Im ε^{-1} – by the energy losses of the strong electrons. Therefore, these optical functions are caused by the unequal number of the valence electrons [12, 13].

Further by the method of the combined diagram Argand the calculated spectra of e_2 and –Im ε^{-1} of HgI_2 crystal were decomposed into the elemental components and following their parameters were obtained: the energies of the maximum E_i and semi-width H_i, amplityds I_i and the areas S_i of bands (Table 1.2).

The decomposed spectra consist the 28 transitions components but the bands № 3, 4, 6, 9, 10, 14, 15, 16, 18, 19, 23 and 24 are not visible in the integral spectra, that is, the decomposition of the initial e_2 and –Im ε^{-1} curves is permitted to establish the 12 disappeared components additional to the 17 maxima of the initial spectra.

It is interestingly to note the parameter S_i, which directly caused the band intensity. The transitions bands mainly divided into two groups: long-wave length (2.3–4.1 eV) and short-wave length (4.2–10 eV) with the S_0 ≈ 150–200 and 10–30 at the deficiencies S_0 to the ~7–10 time. This is evidenced that the intensity of the longitudinal components relatively small in the narrow long-wave length region is raised by the ten with the increased energy.

1.4 CONCLUSION

The spectra of the complex of the fundamental optical functions HgI_2 crystal in the first time were obtained in the energy region 0 to 20 eV for the E⊥c polarization and then the integral spectra of the dielectric permittivity and characteristic energy electron losses were decomposed into the elemental transverse and longitudinal components with their main parameters. The extensive obtained information about the HgI_2 electronic structure assisted to analyze the theoretical calculation of the bands and optical spectra in the wide energy region both qualitatively and in detail.

TABLE 1.2   Parameters of the Decomposed Spectra of the e_2 and $-Im\ \varepsilon^{-1}$ of the HgI_2 Crystal At E⊥c

№	E		H		f (calc)	S	I			S(e_2)/
	e_2	$-Im\varepsilon^{-1}$	e_2	$-Im\varepsilon^{-1}$	e_2	e_2	$-Im\varepsilon^{-1}$	e_2	$-Im\varepsilon^{-1}$	S($-Im\varepsilon^{-1}$)
1	2.339	2.34	0.0017	0.003	0.014	0.11	0.001	42.20	0.320	110
2	2.502	2.49	0.060	0.07	0.024	0.18	0.002	1.90	0.020	90
3	2.630	2.65	0.220	0.19	0.060	0.42	0.003	1.25	0.010	140
4	2.875	2.84	0.200	0.13	0.105	0.68	0.003	2.20	0.017	227
5	3.015	3.00	0.120	0.15	0.078	0.48	0.005	2.60	0.020	96
6	3.115	3.13	0.160	0.12	0.211	1.26	0.007	5.10	0.040	180
7	—	3.23	—	0.12	—	—	0.004	—	0.020	—
8	3.320	3.34	0.185	0.12	0.370	2.07	0.009	7.25	0.050	230
9	3.598	3.58	0.170	0.19	0.247	1.28	0.007	4.87	0.024	183
10	3.738	3.78	0.220	0.18	0.491	2.44	0.011	7.20	0.040	222
11	3.875	3.93	0.200	0.15	0.358	1.72	0.012	5.56	0.050	143
12	4.090	4.15	0.265	0.20	0.757	3.43	0.020	8.42	0.065	172
13	4.275	4.45	0.233	0.28	0.521	2.27	0.040	6.30	0.092	57
14	4.530	—	0.380	—	0.740	3.01	—	5.18	—	—
15	4.950	5.07	0.390	0.35	0.529	1.97	0.089	3.30	0.166	22
16	5.270	5.32	0.130	0.09	0.146	0.52	0.018	2.56	0.130	29
17	5.485	5.48	0.190	0.21	0.157	0.54	0.041	1.82	0.125	13
18	5.668	5.65	0.180	0.22	0.097	0.32	0.031	1.15	0.090	10
19	5.878	5.85	0.170	0.16	0.047	0.15	0.008	0.57	0.030	19
20	6.210	6.22	0.120	0.15	0.043	0.13	0.004	0.70	0.018	33

TABLE 1.2 *(Continued)*

№	E e_2	E $-Im\varepsilon^{-1}$	H e_2	H $-Im\varepsilon^{-1}$	f(calc) e_2	S e_2	S $-Im\varepsilon^{-1}$	I e_2	I $-Im\varepsilon^{-1}$	$S(e_2)/S(-Im\varepsilon^{-1})$
21	6.635	6.67	0.250	0.30	0.372	1.05	0.014	2.70	0.030	75
22	7.010	7.05	0.420	0.30	1.368	3.63	0.014	5.60	0.030	259
23	7.350	7.40	0.380	0.50	0.719	1.82	0.031	3.10	0.040	59
24	7.670	7.78	0.460	0.40	1.095	2.65	0.019	3.74	0.030	139
25	8.120	8.24	0.730	0.71	2.017	4.57	0.052	4.10	0.048	88
26	8.745	9.16	0.860	0.30	2.809	5.90	0.012	4.50	0.025	492
27	9.100	10.10	0.800	1.40	0.846	1.71	0.228	1.40	0.108	8
28	9.650	11.50	0.900	1.65	2.451	4.67	0.347	3.40	0.140	13
Epv1	—	14.20	—	5.60	—	—	3.449	—	0.440	—
Epv2	—	4.74	—	0.40	—	—	0.092	—	0.150	—
Epv2	—	5.07	—	0.35	—	—	0.089	—	0.166	—

ACKNOWLEDGMENTS

This report is accomplished by the program PFFI № 11–02–07038 and 12–02–070007.

KEYWORDS

- Characteristic electron energy losses
- Components
- Dielectric permittivity
- Mercury iodide
- Parameters
- Reflectivity spectra

REFERENCES

1. Gaisler, V. A., Zaletin, V. M., Lijch, N. V., Nojkina, I. N., & Fomin, V. I. (1984). Diiodide Mercuric, Novosibirsk Nauka 104, (In Russian).
2. Ye, J. H., Sherohman, J. M., & Armantrout, G. A. (1976). Theoretical Band Structure Analysis on Possible High–z Detector Materials, IEEE Transactions on Nuclear Sci., V NS, *23(1)*, 117–123.

3. Bulatetcki, K. G., Zaletin, V. M., & Fomin, B. I. (1983). Detectors of the X and γ rays, Pribori and Technika Experimental, (6), 119–122 (In Russian).

4. Sobolev, V.V. (1987). Band and Exciton of the Metal Chalcogenides, Kishinev, Schtiinza, 284p (In Russian)

5. Gros, E. F., & Kaplyanski, A. A. (1955). About the Absorptions Spectra of Some Iodide Crystal, J. Techn, Phys., 55(12), 2061–2068 (In Russian).

6. Sieskind, M., Grun, J. B., & Nikitine, S. (1961). Etude Quantitative Des Specters D' absorption et de Reflexion de Cristaux de HgI₂ Rouge, J. Phys. Rad., T., 22(12), 777–782.

7. Gorbany, I. S., & Rudko, S. N. (1962). The Absorption and Photoluminescence Spectra of HgI₂ Crystals, Optic and Spectr., 12(3), 610–615 (In Russian).

8. Kanzaki, K., & Imai, I. (1972). Optical Spectrum of HgI2, J. Phys. Soc. Japan, 32(4), 1003–1009.

9. Anedda, A., Grilli, E., Guzzi, M., Raga, F., & Serpi, A. (1981). Low Temperature Reflectivity of Red Mercury Iodide, Sol. St. Commun., 39(11), 1121–1123.

10. Chang, Ch. Y., & James, R. B. (1992). Electronic and Optical Properties of HgI₂ Phys Rev B, 46(23), 15040–15045.

11. Ahuja, R., Eriksson, O., Johansson, B., Anluck, S., & Wills, J. M. (1996). Electronic and Optical Properties of Red HgI₂ Phys. Rev. B, 54(15), 10419–10425.

12. Sobolev, V. V., & Nemoshkalenko, V. V. (1988). The Methods of the Calculation Physics in the Theory of Solid State, Electronic Structure of Semiconductors, Kiev, Naukova Dumka, 423p (In Russian).

13. Sobolev, V. V. (2012). Optical Properties and Electronic Structure of Non metals, V. I. Introductions in Theory, Izhevsk, M., IKI, 584p (In Russian).

14. Kalugin, A. I., & Sobolev, V. V. (2005). Electronics Structure of CaF₂ Phys, Rev, B, 71(11), 115112(7).

15. Sobolev, V. V. (2012). Optical Properties and Electronic Structure of Non metals V II, Modeling of the Integral Spectra by the Elemental Bands, Izhevsk, M., IKI, 415p (In Russian).

16. Sobolev, V. V., Sobolev, Val V., & Shushkov, S. V. (2011). Optical Spectra of the Six Silicon Phases, Semiconductors, 45(10), 1247–1250.

17. Anisimov, D. V., Sobolev, V. V., & Sobolev, Val. V. (2012). Optical Properties of in Br in the Region 0 to 30 eV, Proc VIII Intern., Conf., "Amorphous & Microcrystalline Semiconductors," SP Polytech., Inst., 358–359 (In Russian).

18. Timoshkin, A. N., Sobolev, Val V., & Sobolev, V. V. (2000). The Spectra of the Characteristic Electron Energy Losses of the Molibden Dichalcogenides, Phys Solid State, 42(1), 37–40.

19. Sobolev, V. V., & Nemoshkalenko, V. V. (1992). Electronic Structure of Solid State in the Edge of the Fundamental Absorption, V. I. Introduction to the Theory, Kiev, Naukova Dumka, 506p (In Russian).

20. Sobolev, V. V., & Sobolev, Val V. (2012). Electronic Structure of Solid State in the Edge of the Fundamental Absorption, V II, Crystals of II–VI Group, Izhevsk, M. IKI, 607p (In Russian).

CHAPTER 2

EXPERIMENTAL RESEARCH OF GRASSY RAGS IGNITION BY HEATED UP TO HIGH TEMPERATURES CARBON PARTICLE

N. V. BARANOVSKIY and A. V. ZAKHAREVICH

CONTENTS

2.1 INTRODUCTION

Typical enough ignition sources of combustible materials (heated up to high temperatures particles of metals and nonmetals) are described in the domestic and foreign literature [1–5]. Russian works [1–3] are devoted to ignition research of litter of coniferous and deciduous trees by sufficiently small size local heat sources. Results of natural observation of "fire brands" of enough big size formation and their influence on layer of forest fuel [4, 5] are described in foreign literature sources. Researches [6] with samples from cellulose on which the spherical particles of the various sizes made of steel dropped out. It is possible to make a conclusion, that the range of particles of the relative small size is not studied enough. Besides, there are no data about ignition delay times by heated particles for enough widespread types of forest fuel, in particular, grassy rags. In Refs. [4, 5] "fire brands" of any form which were also artificially formed at the special stand are considered [7]. However, observation over control ignitions shows that typical sources of heating are the particles of nonmetals in the form of the small sizes parallelepipeds [1–3].

Research objective of paper is physical modeling of ignition processes of typical forest fuel (grassy rags) by single carbon particle heated up to high temperatures and the basic laws revealing of this phenomenon.

2.2 TECHNIQUE AND OBJECT OF EXPERIMENT

Stand described in details in Refs. [8, 9] was used at present experiment. Ignition delay times were defined by technique [10]. Heating sources of forest fuel ignition were modeled by the graphite bar particles in the form of parallelepipeds with characteristic sizes (peripheral sizes x, y = 8 mm, height h = 15 mm). Experiments were carried out in the range of reference temperatures T_0 from 1113 to 1273 K. This range was chosen to single out the lower limits of ignition of investigated forest fuel by temperature. The choice of carbonaceous particles as heating sources is caused by modeling natural (carrying out of hot particles from fire front on untouched forest fuel layer) and anthropogenous (wood cracking in not extinguished fires) scenarios on influence to forest fuel.

Following technique [11] was used for mathematical processing of experiment results. Average values of ignition delay time of grassy rags by heated particle for each fixed initial temperature of heating source were calculated. Then calculation of root-mean-square deviation and confidential interval by means of Student factors was made.

Research objects were the samples of grassy rags using the packaging density corresponding to a real environment [12]. Materials for the experiment were gathered in spring 2013 on sites along transport and trunk-railways at the borders with large forests (Tomsk area of Tomsk region). Straight before the ignition experiments the samples of grassy rags were dried up in the drying case till full moisture evaporation from the material. The grassy rags has the complex structure including died morphological parts of grassy vegetation, such as stalk, sheet plate and sometimes cereals [12]. All elements from the set of morphological parts with prevalence in a lump of sheet plates were used at formation of samples. Such structure is typical for the grassy rags located in real environment [13].

Scenario of catastrophic forest fire danger [14] when the moisture in forest fuel is absent [15] is considered in experiment. From year to year an urgency of such scenario research increases. For example, catastrophic fire conditions were observed in 2012 in Tomsk region [16] and in 2010 at the Central zone of Russia [17]. There are climate changes in other countries also for which high temperatures and absence of precipitation are typical. For example, fires in 2009 in Australia, 2011 in the USA and 2012 in Southern Europe [18].

2.3 THE BASIC LAWS

Typical video recording captures of grassy rags layer ignition process by heated to high temperatures carbon particle are showed on Fig. 2.1.

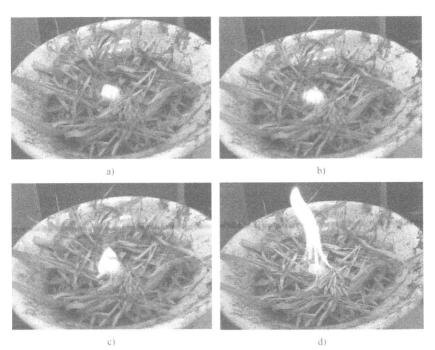

a)
b)
c)
d)

FIGURE 2.1 Typical shots of a video shooting of grassy rags ignition process by carbonaceous particle: inert warming up (t=0.08); having blown gaseous products of pyrolysis in heated area and ignition initiation (t=0.28); formation of a torch of a flame round a particle (t=0.36); growth and spreading of a torch of a flame on forest fuel layer (t=0.48)

Following laws of ignition process are established as result of the analysis of visual observations and the video record. It is necessary to specify, that two variants of interaction of heated particles with forest fuel layer are possible: heating source drops out on a surface of a sheet plate or fails deep into a layer of grassy rags. Penetration of particle into forest fuel layer is characteristic for sources from steel owing to them bigger weights in comparison with particles from carbon. Therefore, we will consider only the mechanism of grassy rags ignition as result of a sheet plate influence to the leaf plate surface. There is an inert warming up of forest fuel layer coming during the first short time period followed by the thermal decomposition of material with release of pyrolysis gaseous products. It is necessary to notice, that release of pyrolysis products has more intensive character, than at pine needles. Forest fuel in surface layer almost completely decomposes with a small amount of carbon residual dropping out to a substrate. There is a transport of gaseous pyrolysis products to a heated layer surface and their mixing with an oxidizer in microporous forest fuel environment. Afterwards the gas mixture is heated with the subsequent stage of ignition. Through the fractions of a second the flame appears at all perimeter of a particle.

Grassy rags ignition delay time dependence from initial temperature of carbonaceous particle is presented in Fig. 2.2. The lower ignition limit for such particle is defined by its initial temperature. Confidential intervals are presented in Fig. 2.2 together with averaged on a series of measurements of forest fuel ignition delay times. High enough values of a confidential interval at initial temperature of carbonaceous particle $T_0 = 1113$ from average values are caused, obviously, by stochastic distribution of separate single blades sheet plates in surface forest fuel layer which directly heated up to high temperature by heated local source. The distance between the separate morphological elements on contact border "forest fuel-heated particle" in the experiments was not fixed to constants. It corresponds to a real arrangement of elements in forest fuel layer.

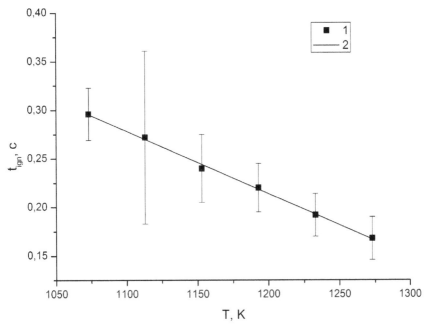

FIGURE 2.2   Dependence of grassy rags ignition delay time on initial temperature of heated particles: 1 – experimental points (average values) with instructions of a confidential interval, 2 – an approximating straight line

Comparative analysis of pine needles and birch leaves ignition time dependences shows that ignition delay time also does not depend from the sizes of ignition source when initial temperature of a particle reaches 1300 K. Dependence of ignition delay time vs. temperature can be approximated a straight line. Similar dependences are obtained and for others forest fuels. Possibly, this is typical for all types of forest fuel. However, it is necessary to notice, there is a complex interconnected diffusive and convective process during the process of forest fuel ignition by local heat source in microporous structure. This results in that the view of ignition delay times dependence of forest fuels considerably differs from typical curves for solid fuel compositions (e.g., [19, 20]).

Also it should be noted that the estimation of conditions and parameters of forest fuel ignition on samples pressed from the dried up grassy rags in the form of discs or plates, should lead to some underestimated comparing to the real, values of delay times and initial temperatures maximum limits of heat sources. Our experiments show the ignition process stability of forest fuel layer very high porosity by a single local heat source. The typical video record (Fig. 2.1) illustrates well the dynamics of studied process.

It is established, that carbonaceous particles interaction with the developed porous surface on forest fuel layer are probable the occurrence of complex physical and chemical transformations of particle frame and products of pyrolysis and combustion of forest fuel

The photo of agricultural fire spreading process near the Chernaya Rechka settlement of Tomsk region is presented on Fig. 2.3 (it is casually fixed by one of paper authors). The burning front didn't have time to grow to big sizes and it was possible to observe products of recreational loading (local population rest) at the back edge of the fire, namely the residue of the charcoal which particles possessed the raised temperature. Carbonaceous particles heated up to high temperature, dropped out on a grassy rags layer, became the reason of its ignition, and convective movements of air weights led to steady burning front spreading and agricultural fire occurrence.

FIGURE 2.3 Agricultural fire on the grassy rags, which has resulted ignition (near settlement Black small river of Tomsk region, May, 2013).

2.4 CONCLUSIONS

Experimental research of ignition processes of typical forest fuel (grassy rags) heated up to high temperatures by nonmetal (carbon) particle is carried out. High probability of forest fuel ignition as a result of local heating source influence is shown. The obtained laws can be used for construction of the generalized forest fuel ignition theory by local sources of heating.

2.5 ACKNOWLEDGMENTS

This work is executed at financial support of the grant of Russian Fond for Basic Research № 12-08-33002 "Development of scientific bases of the general theory of ignition firm, liquid and gelatinous condensed substances at local heating."

KEYWORDS

- **Forest fuel**
- **Heated particle**
- **Ignition**
- **Ignition delay time**
- **Physical modeling**

REFERENCES

1. Zakharevich, A. V., Baranovskiy, N. V., & Maksimov, V. I. (2012). Ignition of Typical Forest Fuel of Deciduous Leaves by Local Energy Source [Zazhiganie Tipichnykh Lesnykh Goryuchikh Materialov Opada Listvennykh Porod Lokalnym Istochnikom Energii], Fire and Explosion safety Pozharovzryvobezopasnost, *21(6)*, 23–28 (In Russian).
2. Zakharevich, A. V., Baranovskiy, N. V., & Maksimov, V. I. (2012). Experimental Research of Ignition Processes of Deciduous Leaves by a Source of Limited Energy Capacity [Eksperimentalnoe Issledovanie Protsessov Zazhiganiya Opada Shirokolistvennykh Porod Derevev Istochnikom Ogranichennoy Energoemkosti], Ecological Systems and Devices Ekologicheskie Sistemy i Pribory, *7*, 18–23 (In Russian).
3. Baranovskiy, N. V., Zakharevich, A. V., & Maksimov, V. I. (2012). Conditions of Forest Fuel Layer Ignition at Local Heating [Usloviya Zazhiganiya Sloya Lesnykh Goryuchikh Materialov Pri Lokalnom Nagreve], Chemical Physics and Mesoscopy Khimicheskaya Fizika i Mezoskopiya, *14(2)*, 175–180 (In Russian).
4. Manzello, S. L., Cleary, T. G., Shields, J. R., & Yang, J. C. (2006). On the Ignitions of Fuel beds by Fire Brands, Fire and Materials, *30*, 77–87.
5. Manzello, S. L., Cleary, T. G., Shields, J. R., & Yang, J. C. (2006). Ignition of Mulch and Grasses by Fire Brands in Wild Land-Urban Interface Fires, International Journal of Wild Land Fire, *15*, 427–431.
6. Hadden, R. M., Scot, S., Lautenberger, Ch, & Fernandez-Pello, C. (2011). Ignition of Combustible Fuel Beds by Hot Particles, an Experimental and Theoretical Study, Fire Technology, *47(2)*, 341–355.
7. Manzello, S. L., Cleary, T. G., Shields, J. R., Maranghides, A., Mel, W., & Yang, J. C. (2008). Experimental Investigation of Firebrands Generation and Ignition of Fuel Beds, Fire Safety Journal, *43*, 226–233.
8. Zakharevich, A. V., Kuznetsov, G. V., & Maksimov, V. I. (2008). Mechanism of Gasoline Ignition by Single Heated up to High Temperatures Metal Particle [Mekhanizm Zazhiganiya Benzina Odinochnoy Nagretoy do Vysokikh Temperature Metallicheskoy Chastitsey], Fire and Explosion Safety Pozharovzryvobezopasnost, *17(5)*, 39–42 (In Russian).

9. Kuznetsov, G. V., Zakharevich, A. V., & Maksimov, V. I. (2008). Ignition of Diesel Fuel by a Single "Hot" Metal Particle [Zazhiganie Dizelnogo Topliva Odinochnoy "Goryachey" Metallicheskoy Chastitsey], Fire and Explosion safety Pozharovzryvobezopasnost, *17(4),* 28–30 (In Russian).

10. Zakharevich, A. V., Kuznetsov, V. T., Kuznetsov, G. V., & Maksimov, V. I. (2008). Ignition of Model Composite Propellants by a Single Particle Heated to High Temperatures, Combustion, Explosion and Shock Waves, *44(5),* 543–546.

11. Gmurman, V. E. (2003). Probability Theory and Mathematical Statistics, Moscow High School, [Teoriya Veroyatnostey i matematicheskaya Statistika, Vyssh Shk, M] 479p (In Russian)

12. Tishkov, A. A. (2003). Fire in Steppes and Savanna [Pozhary v Stepyakh i Savannakh], Questions of Steppe Transaction, Voprosy Stepovedeniya, Institute of Steppes of Ural Branch of the Russian Academy of Sciences, *(4),* 9–22 (In Russian).

13. Marakulina, Yu S., & Degteva, S. V. (2008). Change of Ecological Conditions, Vegetation and Soils at Regenerative Successions on Upland Meadows of the Kirov Region [Izmenenie Ekologicheskikh Usloviy, Rastitelnosti i Pochv Pri Vosstanovitelnykh Suktsessiyakh na Sukhodolnykh Lugakh Kirovskoy Oblasti], Theoretical and Applied Ecology, Teoreticheskaya i Prikladnaya ekologiyan, *2,* 64–73 (In Russian).

14. Flannigan, M. D., Stocks, B. J., & Wotton, B. M. (2000). Climate Change and Forest Fires, Science of the Total Environment, *262(3),* 221–229.

15. Grishin, A. M., Golovanov, A. N., Dolgov, A. A., Loboda, E. L., Baranovskiy, N. V., & Rusakov, S. V. (2002). Experimental and Theoretical Research of Drying of Forest Fuel [Eksperimentalnoe i Teoreticheskoe Issledovanie Sushki Lesnykh Goryuchikh Materialov] Bulletin of TPU Izvestiya, TPU, *305(2),* 31–43 (In Russian).

16. Baranovskiy, N. V. (2012). Experimental Research of forest Fuel Ignition by the Focused Sunlight [Eksperimentalnye Issledovaniya Zazhiganiya Sloya Lesnykh Goryuchikh Materialov Sfokusirovannym Solnechnym Izlucheniem], fire and Explosion Safety Pozharovzryvobezopasnost, *21(9),* 23–27 (In Russian).

17. Devisilov, V. A. (2010). Russian Wood Asks for Mercy and Protection! [Russkiy Less Profit Poshchady i Zashchity!] Safety in a Technosphere, Bezopasnost v Tekhnosfere, *(6),* 3–7 (In Russian).

18. Schmuck, G., San-Miguel-Ayanz, J., Camia, A., Durrant, T., Boca, R., Whitmore, C., Liberta, G., & Corti, P. (2012). Forest Fires in Europe, Middle East and North Africa (2011). JRC Technical Reports Italy, Ispra Joint Research Centre, and Institute for Environment and Sustainability, 109p.

19. Kuznetsov, G. V., Mamontov, Ya G., & Taratushkina, G. V. (2004). Ignition of the Condensed Substance by a "Hot" Particle [Zazhiganie Kondensirovannogo Veshchestva "Goryachey" Chastitsey], Chemical physics Khimicheskaya Fizika, *23(3),* 67–72 (In Russian).

20. Kuznetsov, G. V., Mamontov, Ya G., & Taratushkina, G. V. (2004). Numerical Simulation of Ignition of a Condensed Substance by a Particle Heated to High Temperatures, Combustion, Explosion and Shock Waves, *40(1),* 70–76.

CHAPTER 3

SIMULATION OF STRUCTURAL TRANSFORMATIONS OF SPRING STEEL DURING THE QUENCHING PROCESS

YA. B. BENDERSKIY and K. A. KOPYLOV

CONTENTS

3.1   INTRODUCTION

Now we aware of the different mechanisms of phase and structural transformations occurring during the heat treatment of steels. These mechanisms change the crystal structure, phase composition, morphology, and distributions of structural components, grain size, and other parameters that affect the physical and mechanical properties of the material [1]. Sure, we can get the necessary properties of steels and alloys in several ways:

- *Extensive.* Relies on the creation of new alloys;
- *Intensive.* The basis of this type of classification is a qualitative change in the existing composition of the material during thermal processing by improving heat treatment. In other words, this method is to match the required manufacturing technology and maintains temperature at each process step.

It should be noted that the use of any method of manufacture of steel products require confirmation of physico-mechanical properties. Its main task is to determine the influence of technological factors and different types of processing parameters on the structure. The basis of this method is to assess the effectiveness of the process conditions for heat treatment of steel research results in the collapse of the super-cooled austenite continuous cooling parts from steel and cast iron. Depending on the possible accuracy of the results, this method can be divided into three main groups, each of which requires special training research objects (samples) and a unique instrument technology [2].

As is known from [3], metallographic analysis allows to study the details of the cross sections and fractures semi-finished or finished products. This includes study of the sample after completion of a full cycle of the technological process without the possibility of assessing structural transformation of the material changes in each transaction in real time. The authors noted that the greatest complexity in determining the structure represents the primary measurement of geometrical parameters of structural components. On the one hand, the performance of the metallographic analysis requires great amounts of time and is accompanied by subjective error explorer and, on the other hand, lead to a sample unfit for further use, however, in practice requires avoid these adverse effects, and possible to abandon its use.

For example, in Refs. [4, 5] shows the semi-empirical methods for the calculation of critical cooling rates that can characterize the expected result in advance of formation of structures investigated materials. There are other methods of Refs. [6, 7], where the authors consider two different ways to calculate the phase transformation under cooling at a constant rate. It is also shown that the critical cooling rate is subject to certain functional relationships. Considerable attention is given to justify widespread simple formula of Grange-Kiefer [8] for the upper critical quenching rate v_{KP}:

$$V_{KP} = \frac{T_S - T_m}{\alpha_{\Gamma K} \tau_m},$$ (1)

where T_S – upper temperature limit of transformation; T_m – temperature of "nose" of the curve of ferrite or pearlite; τ_m – the incubation period of the decay at temperature T_m; $\alpha_{\Gamma K}$ – coefficient whose value 1.5.

General method for solving such problems developed by Sheyl [9], and later a similar approach using the graphical integration offered by Steinberg [10, 11].

Known results [12], which describes the use of models to represent the phase transitions steels during thermo-mechanical treatment. The paper describes the changes in the structural condition of the approach of the diffusion boundary with possible redistribution of atoms of carbon and alloying elements. The authors describe an algorithm for solving the problem, however, numerical simulation examples in the process of changing the structure of the phase transitions are not presented.

In Ref. [13], the results of the use of different mathematical models to the study of structural phase transitions occurring on the basic operations of high-temperature thermo-mechanical treatment (HTTT) barrel blanks made of low-alloy steels. Thus, this study indicates the formation of two main structural components of hardened steel with HTTT: marten site and retained austenite. The authors found that the increase in temperature and time release implies the emergence of a list of processes, each of which brings structure to the equilibrium state. This status allows us to give the target material physical and mechanical properties associated with increased reliability, durability, and strength. Ultimately, it is the above-mentioned characteristics led to widespread method HTTT to the manufacturing processes of many engineering products, including springs [14].

Since, in practice often used metallographic analysis to study the properties of materials springs, then to eliminate the influence drawbacks of the method on the final product, we will offer to replace the similar methods of studying the structure of the material on one of the modern tools-ANSYS in the definition and control of non-stationary temperature fields of the spring material for each process operation.

3.2   RESEARCH METHODOLOGY

This paper considers two-dimensional transient cooling process of the spring rod (material-steel 60S2) for each operation of manufacturing method HTTT. Data [15] allow you to specify the physical and mechanical properties of materials. The results of the final structure of the material distribution are determined in real time based on change of temperature of the investigated body.

From Ref. [16], we know the example of calculating the temperature change of the spring at each stage of the manufacturing process by HTTT. As a result, the authors noted a small reduction of temperature after the release of the spring rod from high frequency currents of the inductor – HDTV (12 K in the core rod with the end uncooled during $\tau = 6$) as the assumption is accepted that the temperature field and the structure of the material in this process step is constant.

The next operation process is spiral construction the rod on the mandrel. General scheme of interaction the rod with the mandrel is shown in Fig. 3.1.

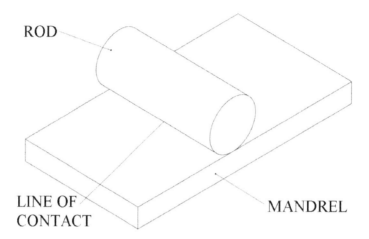

ROD

LINE OF
CONTACT

MANDREL

FIGURE 3.1   Scheme of interactions the rod with the element mandrel.

Enter assumption that during the process operation is carried ideal contact surface of the bar and the mandrel. Two-dimensional transient heat conduction problem was solved. The initial temperature of rod after the inductor HDTV defined as follows:

$$T_{\Pi P}(0; x; y; z) = 1273 \text{ K} \tag{2}$$

Boundary conditions (BC). On the surface of the bar is placed BC of II type:

$$q\big|_S (t; x; y; z) = \varepsilon \cdot S \cdot C \cdot \left[\left(\frac{T_{\Pi OB}}{100} \right)^4 - \left(\frac{T_{OKP}}{100} \right)^4 \right], \tag{3}$$

where $T_{\Pi OB}$ – body surface temperature; T_{OKP} – ambient temperature; $S = f(x; y; z)$ – surface area, which considers the radiation into the environment; ε – emissivity of the surface ($\varepsilon=0.55$ [13]); C – Stefan-Boltzmann constant.

Final operation is the cooling process in the spring sprayer. In this case, on the rod surface defined BC of III type:

$$q\big|_S (t;x;y;z) = \lambda \frac{\partial T}{\partial n}\bigg|_S = \alpha\big[T_{\Pi OB} - T_{OKP}\big], \tag{4}$$

where α – heat transfer coefficient.

In operations winding and cooling on the line of contact with the mandrel rod set BC of IV type:

$$T\big|_{S_1} = T\big|_{S_2} \Rightarrow -\lambda \frac{\partial T_1}{\partial n}\bigg|_{S_1} = -\lambda \frac{\partial T_2}{\partial n}\bigg|_{S_2} \tag{5}$$

The Ref. [17] contains information on the decay of super cooled austenite, representing the corresponding data in the form of diagrams (curves structural transformations). From Ref. [16], assumed that the cooling process of the spring rod with a diameter $d = 0.019$ m occurs when the heat transfer coefficient α=9200 W/(m²×K), process time is $\tau = 5$ s. This condition ensures that the material structure transformation of austenite to marten site when the cooling rate $v_{OXЛ}{\geq}100$ K/s. To represent structural changes over time have developed an algorithm, which allows predicting the change in the structure of the material depending on the unsteady temperature distribution over the cross section.

Methodology provides representation C-shaped curve in the form of polynomial dependencies $T = T(\tau)$ for which was used data from [17]. Distributions of classes of steel are presented in Table 3.1.

TABLE 3.1 Dependence of the Structure of the Material Steel 60S2 from the Cooling Time

τ, с	Temperature T= T(τ), K	Class of steel
$\tau < 4$	$T \geq 573$	Austenite
$\tau < 10$	$T \leq 573$	Martensite
$4 \leq \tau \leq 10$	$T \geq -0.065\tau_3 + 3.078\tau_2 - 39.447\tau + 752.09;$	Ferrite + Karbide
	$T \leq -0.014\tau_4 + 0.738\tau_3 - 12.061\tau_2 + 79.862\tau + 488.69$	

3.3 RESULTS AND DISCUSSION

Figure 3.2 shows the dependence of the temperature distribution (top), distribution structures (bottom) from the time the process of finding a rod on the mandrel.

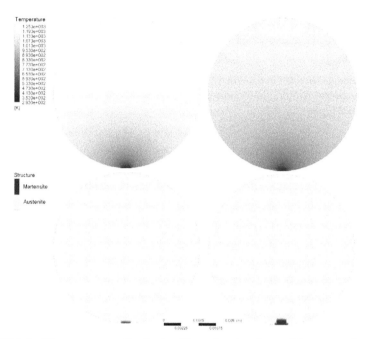

FIGURE 3.2 The temperature distribution (from above) and material structure (bottom) in the rod cross-section, located on a mandrel at $\tau = 2.5$ with the (left), $\tau = 5.0$ with the (right).

The results presented in Fig. 3.2 show that the heat from the mandrel rod causes a change in structure of the material during the process, $\tau = 5$ s and does not affect the core rod, wherein the martensitic structure, and emerging in the area of localized contact.

Cooling process in the spring chamber of spray considered in two formulations:
(a) Without interaction with the mandrel; and
(b) In interaction with the mandrel.

Figure 3.3 shows the temperature change in the core rod (a) on the surface (b), shows a structural transformation temperature of Austenite-Martensite (c), as well as the temperature distribution in the cross section of the structure-forming points on a surface (A) and a core rod (B) versus time for the case of cooling the rod without interaction with the mandrel.

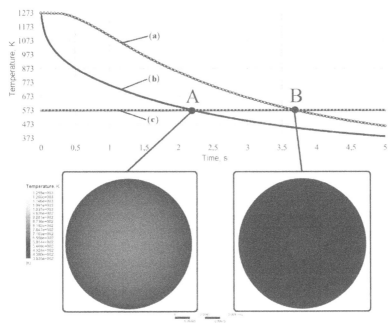

FIGURE 3.3   Changing the temperature in the core rod (a) on a surface (b), the temperature of the structural transformation of Austenite-Martensite (c), the temperature distribution of the bar depending on the process time of cooling in chamber of spray at τ = 2.2 s (left), τ = 3.7 with the (right).

Proceeding from the plot shown in Fig. 3.3, we can say that the structural changes in the material rod should start at τ=2.2 s and ends at τ=3.7 s. This is the fact that during the complete transformation of the Austenite structure to the Martensite will be 1.5 s. Figure 3.4 shows the structure of the material distribution over the cross section of the spring rod. Figure 3.4 also shows that at τ=2.2 s with the material structural changes begin to occur and at τ=3.6 s the material rod has a martens tic structure.

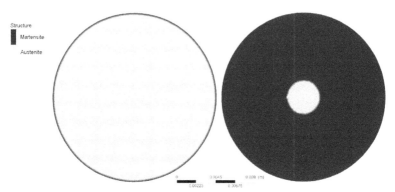

FIGURE 3.4 Distribution over the cross section of the material structure rod depending on the time of the cooling process in the rod in the chamber of spray at τ = 2.2 with (from left), τ = 3.6 s (right).

Figure 3.5 shows the calculated cooling rod in chamber of spray with the interaction with the mandrel.

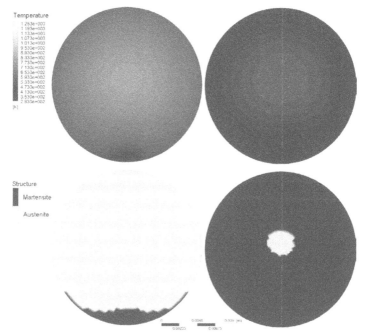

FIGURE 3.5 The distribution of the temperature (from above) and of the structure (bottom) in cross section of the rod, with the mandrel located in the chamber of spray at τ = 2.0 (left), τ = 3.4 (right).

Offset areas seen more ductile structures from the center bar to the top of the spring section, and, in accordance with the cooling conditions during the period of time τ=3.5 s occurs with full harden ability material and changing the structure of austenitic to martens tic.

3.4   CONCLUSION

The dependence of unsteady temperature field and the structure of the material in the cross section of the spring rod was developed. Calculation for each process step of manufacturing springs HTTT method was implemented. Findings:

1. Occurrence of a local region in the formation of the martensitic structure during winding onto the mandrel bar at the contact portion;
2. Full-time transformation austenite to marten site structure with uniform cooling rod in chamber of spray without heat due to the interaction with the mandrel is τ=1,5 s.
3. Offset areas more ductile structures (austenite) from the center of the bar to the top of the spring section with cooling mandrel rod in sprayer.

As consideration for the work provided for a number of assumptions of the mathematical model, including the lack of heat transfer between the fluid (water) and heated metal surface, evaporation is not considered, so further work is supposed to combine the solution of conjugate heat transfer, and structural phase transitions of substances.

KEYWORDS

- **Spring steel**
- **Structural transformations**
- **The method of calculation**
- **The spring**
- **The temperature distribution**

REFERENCES

1. Gun, P. S., Shubin, I. G., & Sokolov, A. A. (2005). Kachestvo Djubelja i Tehnologija Ego Proizvodstva, Vestnik MGTU, *1*, 58–63.
2. Sergeev, Ju G., Stoljarova, N. A., & Kislenkov, V. V. (2003). Material Ovedenie, Metallograficheskij Analiz i Diagrammy Sostojanija, Uchebnoe Posobie, SPb, SPbGPU, 80 s.
3. Bogomolova, N. A. (1982). Prakticheskaja Metallografija, Uchebnik Dlja Tehn., Uchilishh 2-e izd., Ispr., M, Vyssh shkola, 272 s.
4. Blanter, M. E. (1984). Teorija Termicheskoj Obrabotki, M, Metallurgija, 328 s.
5. Kachanov, N. N. (1978). Prokalivaemost' Stali M, Metallurgija, 192 s.

6. Mirzaev, D. A., & Okishev, Ju K. (2002). Obrazovanie Ferrita v Staljah, Fazovye i Strukturnye Prevrashhenija v Staljah, *2*, 86120.

7. Mirzaev, D. A., Okishev, Ju K, & Mirzaeva, K. D. (2002). Prevrashhenie Austenita Stalej v Uslovijah Nepreryvnogo Ohlazhdenija, izvestija Cheljabinskogo Nauchnogo Centra UrO RAN, *4*, 21–30.

8. Grange, R. A., & Kiefer, J. M. (1941). Transformation of Austenite on Continuous Cooling and Its Relation to Transformation at Constant Temperature, Trans ASM, *29(1)*, 85–114.

9. Scheil, E. (1934). Anlaufzeit Der Austenitumwandlung, Arch., Eisenhüttenwesen, 35, 8, 12P, 565–567.

10. Shtejnberg, S. S. O. (1937). Zavisimosti Mezhdu Skorost'ju Ohlazhdenija, Skorost'ju Prevrashhenija, Stepen'ju Pereohlazhdenija i Kriticheskoj Skorost'ju Zakalki, Trudy UFAN, 9–11.

11. Shevjakina, L. E. (1950). Svjaz' Mezhdu Protekaniem Prevrashhenija Austenita Pri Nepreryvnom Ohlazhdenii i Dannymi Izotermicheskoj Diagrammy, Fazovye Prevrashhenija v Zhelezouglerodistyh Splavah, 101–120.

12. Trusov, P. V., & Shvejkin, A. I. (2012). Mnogourovnevye Modeli Neuprugogo, Deformirovanija Materialov i Ih Primenenie Dlja Opisanija Jevoljucii Vnutrennej Struktury, Fizicheskaja Mezomehanika, *1*, 33–56.

13. Mujzemnek, Ju A. (2007). Matematicheskoe Modelirovanie Strukturno-Fazovyh, Prevrashhenij v Processah Vysokotemperaturnoj Termomehanicheskoj Obrabotki, Vestnik Izhevskogo Gosudarstvennogo Tehnicheskogo Universiteta, *4*, 101–105.

14. Chetkarev, V. A., Dement'ev, V. B., & Shavrin, O. I. (1996). Analiz i Optimizacija Tehnologij Uprochnenija Metalloprodukcii Metodom VTMO, Izhevsk, IPM UrO RAN, 136 s.

15. Rajces, V. B. (1980). Termicheskaja Obrabotka V, Pomoshh' Rabochemu-Termistu, M, Mashinostroenie, 192 s.

16. Benderskij, Ja B., & Kopylov, K. A. (2011). Modelirovanie Processov Teploobmena Pri Izgotovlenii Pruzhin Metodom Vysokotemperaturnoj Termomehanicheskoj Obrabotki (VTMO), Himicheskaja Fizika i Mezoskopija, *1*, 28–36.

17. Popov, A. A., & Popova, L. E. (1965). Izotermicheskie i Termo kineticheskie Diagrammy Raspada Pereohlazhdennogo Austenita, M, Metallurgija, 496 s.

CHAPTER 4

DEFINITION OF THERMOCHEMICAL CHARACTERISTICS FOR DISPERSED CONDENSED SUBSTANCE IGNITION IN THE CONDITIONS OF LOCAL ENERGY SUPPLY

D. O. GLUSHKOV, A. V. ZAKHAREVICH, and P. A. STRIZHAK

CONTENTS

4.1   INTRODUCTION

Investigation of conditions and characteristics of condensed substances (CS) ignition such as composite propellants [1–5], combustible forest materials [6–10] and liquid fuels [11–15] by local power sources is an actual problem. On the one hand, this method of fuel ignition is the most promising for energy efficiency of combustion activation in special and power plants. On the other hand, an interaction between high energy substances and hot particles generated by friction or impacts is characterized by high risk at fuel transportation and storage [14].

Experimental investigations of integrated characteristics and behavior of heat and mass transfer under CS ignition by single hot particles is not always possible. The reasons for this are the process specificity (power source temperature changes during the induction period) and the limited facilities of experimental methods [3, 6–8, 14]. Mathematical simulation allows excepting a number of problems connected, for example, with registration of parameters of fast-developing physical process as well as lets to explore successive steps from initial moments of porous CS warming-up till its ignition in more details. But numerical research of ignition conditions and combustible substance characteristics is possible under known kinetic parameters of the exothermal reaction. Determination values of the activation energy and preexponential factor by Arrhenius dependence of the oxidation reaction rate on ignition source's temperature [16] is a rather complicated task especially for the experiments with dispersed CS. It's conditioned by the fact that the heat exchange conditions for a single particle of a milled material and a massive heating source (such as metal plate with constant surface temperature) differ essentially from heating conditions for dispersed CS layer by local source (such as a single metal particle warmed to high temperatures [1–4, 6–9, 11–15]). Therefore, the information about these characteristics is not yet available even for common porous combustible substances.

The purpose of this work is determination of thermo chemical characteristics (activation energy and pre exponential factor) for reaction of gas-phase dispersed condensed substance ignition (on the example of coal dust particles) according to experimental studies results.

4.2   PROBLEM STATEMENT

Investigation of physical and chemical processes taking place during the ignition has been realized in system "dispersed CS-energy source with limited heat content-air." A scheme of the solution domain of the problem is represented in Fig. 4.1.

It was assumed that the local energy source-the small-size ($r_p = r_1 = 3 \times 10^{-3}$ m, $h_p = z_2 - z_1 = 3 \times 10^{-3}$ m) disk-shaped steel particle heated to high temperature is deposited on the surface of the porous CS. As a result of warming up in near-surface layer of substance its thermal destruction process is initiated close to the lower boundary of power source. Gaseous pyrolysis products are filtered freely inside the warmed-up layer under the heating source to the border "substance-air" due to the porosity of the dispersed CS layer and pass to oxidant area. The gaseous thermal decomposition products are mixed with air due to the diffusion mass transfer. Gas mixture is formed. The following warming-up of the combustible gas-oxidant mixture is caused by its movement along the vertical borders $r=r_1$ (Fig. 4.1) of the "hot" particle. An oxidation process becomes irreversible when the gas mixture temperature and the concentration of the gaseous combustible components take sufficient values. Gas-phase ignition occurs.

4.3 EXPERIMENTAL TECHNIQUE

The experimental investigations of ignition for pored CS by the single steel particle warmed to high temperatures was held used the plant according the methods [3].

Analysis and generalization of experimental data [3, 6–8, 14] on dependence of ignition delay time on initial temperature of heating source allow formulating an important statement regulating experimental conditions for determination of kinetic parameters. It is reasonable to conduct an experiment at the maximum possible initial temperature (T_p) of the particle (source of energy) for ensuring minimal errors of ignition delay time (t_d) measurements. For example, the random errors of t_d measurement reach a minimum (less than 7%) at $T_p > 1050$ K. Also the thickness (h_p) of the particle has no effect on t_d at high T_p. The deviations in delay time of substance ignition by particles of different thickness reach 300% [3, 6–8, 14] at relatively low values of T_p. Therefore, the heating of the steel particle was executed to the temperatures of more than 1050 K. The experimental results are shown in Fig. 4.2. The kinetic parameter values of ignition reaction–E_1 (activation energy) and the product $Q_0 k_1^\circ$ (a thermal effect of reaction and pre exponential factor, respectively) were calculated by the expression [16]:

$$\cdot t_d = 1,18\left(1 - \frac{T_0}{T_p}\right)\sqrt{T_p - T_0}\,\frac{C}{Q_0 k_1^0}\sqrt{\frac{E_1}{R_t}}\,\exp\left(\frac{E_1}{R_t T_p}\right) \tag{1}$$

Two points of the experimental curve (Fig. 4.2, curve 1) were used for calculating of two unknowns. The points make a connection between T_p and t_d in area of relatively high initial temperatures of energy source. Thus it is sufficiently to solve a system of two transcendental equations for determination the kinetic parameters.

The kinetic parameters values of the oxidation reaction were calculated for the variation range of the hot particle initial temperature from 1050 K to 1200 K: $E_1 = 84 \times 10^3$ J/mol, $Q_0 k_1^\circ = 5.516 \times 10^{11}$ J/(kg $\times$ s).

4.4   MATHEMATICAL MODEL

Mathematical simulation has been executed in order to analyze the reliability of the determined values of the kinetic ignition parameters by means of Arrhenius dependence (Fig. 4.2, curve 1) of the rate of oxidation reaction on the local energy source's initial temperature.

Numerical simulation has been held with the following assumptions:

1.   The substance with known thermal characteristics is appeared as a result of thermal destruction of the dispersed CS.
2.   A layer of the substance adjacent to the surface is in an unstrained state and there is no deformation of the surface after the fallout of the particle (the particle is not embedded in the surface layer).
3.   Possible processes of particle's burnout and crystallization of ignition source are not considered.
4.   A gas-arrival from the surface area of dispersed CS closed by the particle ($z=z_1$, $0<r<r_1$) is distributed in close proximity to the heating source.

The following complex of ignition criteria was used which allows considering that specific features of the heat-and-mass transfer processes in the system (Fig. 4.1) with the local power source during the induction period:

1.   Energy released by the oxidation reaction of the gaseous products of thermal decomposition of dispersed CS's particles is greater than heat transferred from the heating source to the reaction zone.
2.   The gas mixture temperature is greater than the initial temperature of the hot particle in the reaction zone of intensive oxidation.

The ignition problem has been solved in an axially symmetric formulation in a cylindrical coordinate system with the origin coincident with the symmetry axis of the hot particle. The given below system of nonlinear non-stationary differential equations describes a complex of the heat and mass transfer processes with the phase transformations and chemical reactions at $0<t<t_d$ (Fig. 4.1). It was assumed that diffusion is the main mechanism for transport of combustible gas into reaction zone. Possible convective mass transfer processes were ignored in the solution area. It can be explain that the intensity of blowing of gaseous thermal decomposition products in the boundary area of a local energy source (Fig. 4.1) is less than mass velocity of vapor generated by liquid fuel heating [11–13] under otherwise equal conditions.

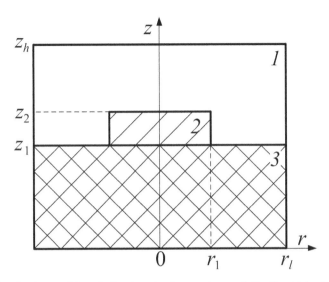

FIGURE 4.1 A scheme of the solution domain: *1*–gas mixture, *2*–"hot" particle, *3*–dispersed condensed substance.

The energy equation for the gas mixture of oxidant (air) with gas components of porous CS particles thermal decomposition ($r_1 < r < r_l, z_1 < z < z_2; 0 < r < r_l, z_2 < z < z_h$):

$$\rho_1 C_1 \frac{\partial T_1}{\partial t} = \lambda_1 \left(\frac{\partial^2 T_1}{\partial r^2} + \frac{1}{r} \frac{\partial T_1}{\partial r} + \frac{\partial^2 T_1}{\partial z^2} \right) + Q_0 W_0, \tag{2}$$

where $W_0 = \rho_1 C_0^1 C_f^1 k_1^0 \exp \left(-\dfrac{E_1}{R_t T_1} \right)$ is mass rate of combustible gases oxidation in the air [17].

The heat transfer equation for the metallic particle ($0 < r < r_1, z_1 < z < z_2$):

$$\rho_2 C_2 \frac{\partial T_2}{\partial t} = \lambda_2 \left(\frac{\partial^2 T_2}{\partial r^2} + \frac{1}{r} \frac{\partial T_2}{\partial r} + \frac{\partial^2 T_2}{\partial z^2} \right) \tag{3}$$

The energy equation for porous CS ($0 < r < r_l, 0 < z < z_1$):

$$\rho_3 C_3 \frac{\partial T_3}{\partial t} = \lambda_3 \left(\frac{\partial^2 T_3}{\partial r^2} + \frac{1}{r} \frac{\partial T_3}{\partial r} + \frac{\partial^2 T_3}{\partial z^2} \right) + Q_3 W_3, \tag{4}$$

where $W_3 = \varphi_3 \rho_3 k_3^0 \exp \left(-\dfrac{E_3}{R_t T_3} \right)$ is mass rate of porous CS particles thermal decomposition [17].

The diffusion equation for the combustible gases in the oxidant (air) ($r_1 < r < r_l$, $z_1 < z < z_2$; $0 < r < r_l, z_2 < z < z_h$):

$$\rho_f^1 \frac{\partial C_f^1}{\partial t} = \rho_f^1 D_f^1 \left(\frac{\partial^2 C_f^1}{\partial r^2} + \frac{1}{r} \frac{\partial C_f^1}{\partial r} + \frac{\partial^2 C_f^1}{\partial z^2} \right) + W_0 \tag{5}$$

The balance equation for the gas mixture ($r_1 < r < r_l, z_1 < z < z_2$; $0 < r < r_l$, $z_2 < z < z_h$):

$$C_f^1 + C_O^1 = 1 \tag{6}$$

The effective thermo physical characteristics of porous CS:

$$\lambda_3 = \lambda_f^3 \varphi_f^3 + \lambda_O^1 \varphi_O^3,$$

$$C_3 = C_f^3 \varphi_f^3 + C_O^1 \varphi_O^3,$$

$$\rho_3 = \rho_f^3 \varphi_f^3 + \rho_O^1 \varphi_O^3,$$

where φ_f^3 is dimensionless volume share of porous CS particles; φ_O^3 is dimensionless volume share of oxidant (air).

The dimensionless volumes of gas mixture components (components of CS particles thermal decomposition and air) were determined by it dimensionless mass concentration in gas mix:

$$\varphi_f^1 = \frac{C_f^1 / \rho_f^1}{C_f^1 / \rho_f^1 + C_O^1 / \rho_O^1},$$

$$\varphi_f^1 + \varphi_O^1 = 1$$

The thermo physical characteristics of gas mixture (oxidant (air) and gas components of CS thermal decomposition):

$$\lambda_1 = \lambda_f^1 \varphi_f^1 + \lambda_O^1 \varphi_O^1,$$

$$C_1 = C_f^1 \varphi_f^1 + C_O^1 \varphi_O^1,$$

$$\rho_1 = \rho_f^1 \varphi_f^1 + \rho_O^1 \varphi_O^1,$$

where φ_f^1 is dimensionless volume of combustible gases; φ_O^1 is dimensionless volume of oxidant (air).

Nomenclature were accepted: t_d is ignition delay time, s; r, z is coordinates of the rectangular system, m; r_1, z_h is solution domain extent, m; T_0 is initial temperature of air and CS, K; T_p is initial temperature of "hot" particle, K; λ is thermal conductivity, $W/(m \times K)$; ρ is density, kg/m^3; C is specific heat, $J/(kg \times K)$; Q_o is thermal effect of oxidation reaction, J/kg; Q_3 is thermal effect of gasification reaction, J/kg; k_1°, k_3° is preexponential factors, s^{-1}; E_1 is activation energy of oxidation reaction, J/mol; E_3 is activation energy of gasification reaction, J/mol; R_t – absolute gas constant, $J/(mol \times K)$; C_f^1 is dimensionless mass concentration of combustible gases in the gas mixture ($0 < C_f^1 < 1$), C_o^1 is dimensionless mass concentration of oxidant (air) in the gas mixture; subscripts "1," "2," "3" correspond to gas mixture, steel particle, dispersed condensed substance.

Regional conditions for an ignition problem (Fig. 4.1) have the following appearance.

Initial conditions at $t=0$:

$T_1 = T_0$, $C_f^1 = 0$; $r_1 < r < r_1$, $z_1 < z < z_2$; $0 < r < r_1$, $z_2 < z < z_h$;
$T_2 = T_p$, $0 < r < r_1$, $z_1 < z < z_2$;
$T_3 = T_0$, $\phi_f^3 = \phi_f^{3o}$; $0 < r < r_1$, $0 < z < z_1$.

Boundary conditions at $0 < t < t_d$:

1. On symmetry axis and external borders we set for all equations the condition of gradients vanishing of corresponding functions:

   $r=0$, $0 < z < z_1$; $r=r_1$, $0 < z < z_1$: $\dfrac{\partial T_3}{\partial r} = 0$;

   $r=0$, $z_1 < z < z_2$: $\dfrac{\partial T_2}{\partial r} = 0$;

   $r=0$, $z_2 < z < z_h$: $\dfrac{\partial T_1}{\partial r} = 0$, $\dfrac{\partial C_f^1}{\partial r} = 0$;

   $r=r_1$, $z_1 < z < z_h$: $\dfrac{\partial T_1}{\partial r} = 0$, $\dfrac{\partial C_f^1}{\partial r} = 0$;

   $z=0$, $0 < r < r_1$: $\dfrac{\partial T_3}{\partial z} = 0$;

   $z=z_h$, $0 < r < r_1$: $\dfrac{\partial T_1}{\partial z} = 0$; $\dfrac{\partial C_f^1}{\partial z} = 0$.

2. Thermal interaction between system components (Fig. 4.1) was described by boundary condition of type IV with allowance for porous CS particles gasification on the site of a surface closed by "hot" metallic particle:

   $r=r_1$, $z_1 < z < z_2$: $-\lambda_2 \dfrac{\partial T_2}{\partial r} = -\lambda_1 \dfrac{\partial T_1}{\partial r}$, $T_2 = T_1$, $\dfrac{\partial C_f^1}{\partial r} = 0$;

   $z=z_2$, $0 < r < r_1$: $-\lambda_2 \dfrac{\partial T_2}{\partial z} = -\lambda_1 \dfrac{\partial T_1}{\partial z}$, $T_2 = T_1$, $\dfrac{\partial C_f^1}{\partial z} = 0$;

   $z=z_1$, $0 < r < r_1$: $-\lambda_3 \dfrac{\partial T_3}{\partial z} = -\lambda_2 \dfrac{\partial T_2}{\partial z}$, $T_3 = T_2$, $\dfrac{\partial C_f^1}{\partial z} = 0$;

   $z=z_1$, $r_1 < r < r_1$: $-\lambda_3 \dfrac{\partial T_3}{\partial z} = -\lambda_1 \dfrac{\partial T_1}{\partial z}$, $T_3 = T_1$, $-\rho_1 D_f^1 \dfrac{\partial C_f^1}{\partial z} = W_g$,

   where $W_g = \displaystyle\int\limits_{z=0}^{z=z_1} \phi_3 \rho_3 k_3^0 \exp\left(-\dfrac{E_3}{R_t T_3}\right) dz$

Gasification rate of dispersed CS particles on the surface closed by a "hot" particle was defined by a ratio:

$$z = z_1, \ 0 < r < r_1: \ W_{g\Sigma} = \int_{r=0}^{r=r_1} W_g(r)dr$$

Total gas-arrival $W_{g\Sigma}$ was distributed in a small vicinity of a particle according to expression:

$$z = z_1, \ r_1 < r < r_1 + 10h_r: \ W_g^S(r)_k = W_g(r)_k + \frac{1,1 - 0,1k}{10} W_{g\Sigma}, \ k = 1; \ 2; \ldots; \ 10,$$

where $W_g(r)_k$ is mass rate of gasification on k-th step along an axis r without additional gas-arrival from the site of a surface closed by a "hot" particle; $W_g^S(r)_k$ is total gasification rate on k-th step along an axis r in a vicinity of a "hot" particle.

The set of nonlinear non-stationary differential equations in private derivatives (2)–(6) with the corresponding initial and boundary conditions was solved by the method of final differences with use of algorithms [11–13].

4.5   RESULTS AND DISCUSSION

The numerical investigation was carried out for the following values of parameters: CS and air initial temperature $T_0 = 300$ K, steel particle initial temperature $T_p = 1050 \div 1200$ K; volume fraction of a substance capable of chemical reaction $\phi_3 = 0.5$; kinetic parameters of the thermal decomposition reaction of CS–$E_3 = 195 \times 10^3$ J/mol, $Q_3 k_3^\circ = 25.5 \times 10^{14}$ J/(kg $\times$ s); kinetic parameters of the oxidation reaction of the gas mixture were calculated by the experimental dependence $t_d = f(T_p)$:

$$E_1 = 84 \times 10^3 \text{ J/mol}, \ Q_0 k_1^\circ = 5.516 \times 10^{11} \text{ J/(kg} \times \text{s).} \tag{7}$$

Thermo physical characteristics of substances are listed below [18–20]:

$$\lambda_0^1 = 0.026 \text{ W/(m} \times \text{K)}; \ \rho_0^1 = 1.161 \text{ kg/m}^3; \ C_0^1 = 1190 \text{ J/(kg} \times \text{K)};$$

$$\lambda_2 = 49 \text{ W/(m} \times \text{K)}; \ \rho_2 = 7831 \text{ kg/m}^3; \ C_2 = 470 \text{ J/(kg} \times \text{K)};$$

$$\lambda_f^1 = 0.072 \text{ W/(m} \times \text{K)}; \ \rho_f^1 = 2.378 \text{ kg/m}^3; \ C_f^1 = 3876 \text{ J/(kg} \times \text{K)};$$

$$\lambda_f^3 = 0.186 \text{ W/(m} \times \text{K)}; \ \rho_f^3 = 1400 \text{ kg/m}^3; \ C_f^3 = 1310 \text{ J/(kg} \times \text{K)}.$$

Has been found (see Fig. 4.2) that the experimental (curve 1) and calculated (curve 2) as a result of solving the problem (2–6) values of t_d (for identical temperatures) differ nearly by 25% providing by relatively low values of T_p. Probably the reasons can be explained by the features of gas-phase ignition with local heating of substance as well as by the specificity of heat and mass transfer in the heated layer of porous CS.

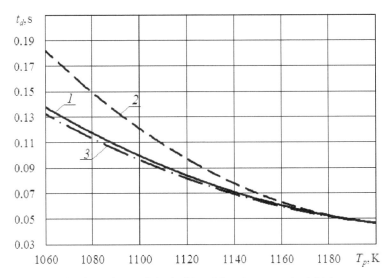

FIGURE 4.2 The dependence of the ignition delay time t_d on the initial temperature T_p of "hot" particle: *1*–experimental curve, *2*–theoretical curve at $Q_o k_1^\circ$=const, *3*–theoretical curve at $Q_o k_1^\circ = f(T_p)$.

For a formal description of the ignition process the analysis of received experimental data has been executed considering a possible dependence between the pre-exponential factor and temperature. The following approximation expression was received:

$$Q_o k_1^\circ = (2.193 \times 10^6 - 2.493 \times 10^3 \times T_p + 0.987 \times T_p^2) \text{ J/(kg} \times \text{s) at } T_p = 1050 \div 1200 \text{ K.} \quad (8)$$

The Fig. 4.2 shows the theoretical dependence $t_d = f(T_p)$–curve 3 obtained by numerical simulation of the ignition process using the characteristics of Eq. (8). An acceptable agreement between the experimental and calculated values of the ignition delay times at low temperatures is can be noted. The extreme deviations of the temperature do not exceed 3.5%.

The type of the approximation dependence of Eq. (8) allows to suggest that the model being a basis of Eq. (1) gives a very good description of the actual process due to some "redundancy" of heat entering to the reaction zone. The heat transfer conditions in the case of dispersed substance as well as for the mixture of fuel and oxidant have a little effect on the chemical reaction rate if T_p is greater than some limiting value (in that case $T_p \approx 1180$ K). At relatively low temperature of local source the conditions of energy supply to the zone of intensive interactions have a great importance. Summarizing the results of the research it can be concluded that we can always choose such variation range of the initial temperature of the heat source where the kinetics of the ignition process can be determined by the formula (1) [16] with high accuracy for the ignition of dispersed CS at local heating conditions. At the same time, the dependence of Eq. (8) is an illustration, most probably, that at low (relatively) source temperatures (especially near the critical values) the details of the heat transfer process are much more important than a range of high temperatures. Accordingly, the values of activation energy and preexponential factor (7) can be considered as true values, and the Eq. (8) can be regarded as an approximation, which can be used in practice in the whole range of temperature (Fig. 4.2) of the local heating source.

More complex schemes of thermokinetic interaction between pyrolysis gases of the condensed substance particles and an oxidant are associated with the necessity to determine a large number of thermokinetic constants for reacting fuel components at each stage of the reaction. However, such approach leads to significant complication of the analysis of conditions and ignition characteristics for particular porous condensed substances.

4.6 CONCLUSION

Investigation results allow making a conclusion about the possibility to determine the kinetic characteristics of the gas-phase ignition of dispersed CS by a local source using relatively simple approach developed long ago [16].

ACKNOWLEDGMENTS

This work was supported by the Ministry of Education and Science of the Russian Federation (No. 14.B37.21.2071) "Modeling of physical processes in elements of the power effective heat power equipment."

KEYWORDS

- **Dispersed condensed substance**
- **Ignition**
- **Local heating**
- **Thermo chemical characteristics**

REFERENCES

1. Kuznetsov, G. V., Mamontov, G. Y., & Taratushkina, G. V. (2004). Ignition of Condensed Matter by the "Hot" Particle, Russian Journal of Physical Chemistry, B, *23(3)*, 67–72.

2. Glushkov, D. O., & Strizhak, P. A. (2012). The Influence of a Local Energy Source Configuration on Ignition Conditions of Structurally Non-uniform Solid Condensed Substance, Physical Chemistry and Mezoskopiya, *14(3)*, 334–340 (in Russian).

3. Zakharevich, A. V., Kuznetsov, G. V., Maksimov, V. I., & Kuznetsov, V. T. (2008). Ignition of Model Composite Propellants by a Single Particle Heated to High Temperatures, Combustion, Explosion, and Shock Waves, *44(5)*, 543–546.

4. Burkina, R. S., & Mikova, E. A. (2009). High-Temperature Ignition of a Reactive Material by a Hot Inert Particle with a Finite Heat Reserve, Combustion, Explosion, and Shock Waves, *45(2)*, 144–150.

5. Arkhipov, V. A., & Korotkih, A. G. (2011). Features of Ignition and Thermal Decomposition of HEM on the Basis of Ammonium Nitrate and the Active Binding, Physical Chemistry and Mezoskopiya, *13(2)*, 155–164 (in Russian).

6. Zakharevich, A. V., Baranovsky, N. V., & Maksimov, V. I. (2012). Ignition of Combustible Forest Materials by the Single Particles Heated to High Temperatures, Fire and Explosion Safety, *21(4)*, 13–16 (in Russian).

7. Kuznetsov, G. V., Zakharevich, A. V., Maksimov, V. I., & Moshkov, A. G. (2012). Ignition Conditions of Wood working Waste, Fire and Explosion Safety, *21(5)*, 21–23 (in Russian).

8. Zakharevich, A. V., Baranovsky, N. V., & Maksimov, V. I. (2012). Ignitions of Typical Combustible Forest Materials of deciduous Breeds by Local Power Source, Fire and Explosion Safety, *21(6)*, 23–28 (in Russian).

9. Kulesh, R. N., & Subbotin, A. N. (2009). Peat Ignition by an External Local sources of Heat, Fire and Explosion Safety, *18(4)*, 13–18 (in Russian).

10. Grishin, A. M., Golovanov, A. N., & Medvedev, V. V. (1999). On the Ignition of a Layer of Combustible Forest Materials by Light Radiation, Combustion, Explosion, and Shock Waves, *35(6)*, 618–621.

11. Kuznetsov, G. V., & Strizhak, P. A. (2009). Simulations of the Ignition of a Liquid Fuel with a Hot Particle, Russian Journal of Physical Chemistry, B, *3(3)*, 441–447.

12. Kuznetsov, G. V., & Strizhak, P. A. (2009). 3D Problem of Heat and Mass transfer at the Ignition of a Combustible Liquid by a Heated Metal Particle, Journal of Engineering Thermo physics, *17(1)*, 72–79.

13. Kuznetsov, G. V., & Strizhak, P. A. (2010). Transient Heat and Mass Transfer at the Ignition of Vapor and Gas Mixture by a Moving Hot Particle, International Journal of Heat and Mass Transfer, *53(5, 6)*, 923–930.

14. Zakharevich, A. V., Kuznetsov, G. V., Maksimov, V. I., Panin, V. F. & Ravdin, D. S. (2008). Assessment of Fuel Oil Fire Danger at the Conditions of Load, Storage and Transport at Thermal Power Plants, Bulletin of the Tomsk Polytechnic University, *313(4)*, 25–28 (in Russian).

15. Burkin, V. V., & Burkina, R. S. (2006). Comparison of Condensed Charges Ignition Parameters at the Thermo Chemical and Electro Plasma Influences, Physical Chemistry and Mezoskopiya, *8(1)*, 104–113 (in Russian).

16. Vilyunov, V. N., & Zarko, V. E. (1989). Ignition of Solids, Amsterdam, Elsevier Science Publishers.

17. Frank, Kamenetsky, D. A. (1969). Diffusions and heat Transfer in Chemical Kinetics, New York Plenum Press.

18. Yurenev, V. N., & Lebedev, P. D. (1975). References Book on Thermo physical Parameters, Moscow, Eergiya, *1* (in Russian).

19. Yurenev, V. N., & Lebedev, P. D. (1975). References Book on Thermo physical Parameters, Moscow Eergiya, *2* (in Russian).

20. Vargaftik, N.B. (2006). References Book on Thermo physical Parameters of Gases and Liquids, Moscow Stars (in Russian).

CHAPTER 5

INFLUENCE OF ELECTRON BEAM PARAMETERS ON IGNITION OF ENERGETIC MATERIALS

G. A. IVANOV and A. V. KHANEFT

CONTENTS

5.1   INTRODUCTION

Recently the initiation of organic explosives by nanosecond electron pulses has been studied actively [1–5]. The most examined is PETN ($C_5H_8N_4O_{12}$). The critical energy density of an electron beam $W*$, which leads to the initiation of PETN in the area of the absorption of the electron beam, is 15 J/cm² at the initial energy of electrons $E_0 = 250$ keV. The time of delay at the current energy density is around 3.45 ms [2]. The energy, released in the area of electron beam absorption, is not enough for the detonation of the rest of a sample. The detonation of PETN takes place while the absorption of the electron beam, which energy density is $W_D \sim 60$ J/cm² and $E_0 = 450$ keV [3]. The detonation of PETN, glued to a copper plate, occurs at $W* \sim 15$ J/cm² and electron energy $E_0 = 250$ keV. [4]. The detonation spreads from the boundary "explosive metal" to the area of electron beam absorption. A negative charge is known to induce a positive one in a subsurface layer. That causes mirror force, which naturally accelerates electrons. Nowadays there are two points of view on the mechanism of PETN initiation by the electron beam: the electric discharge model [5] and the thermal model [2, 6]. Thus, If PETN detonation can be caused by electric breakdown then the direction of its propagation contradicts to the experiment described in article [4].

The work of Ref. [6] examines the thermal model of PETN initiation, which has the system of coupled thermo elasticity equations as its basis. The research takes into account the dependence of chemical reaction energy on elastic stress. The calculation results qualitatively correspond to the experimental data [2] on the delay time of PETN initiation by the electron beam. The works of Refs. [7, 8] analyze radiation-thermal mechanism of PETN initiation by the beam of nanosecond electrons. The mechanism has the system of coupled thermo elasticity equations completed with the autocatalysis equation as its basis. The introduction of the stage of autocatalysis allows to decrease the time of PETN initiation delay in the model.

The objective of the current paper is to find out if the thermal model qualitatively describes the experimental data on PETN initiation by a broad electron beam ignoring thermo elastic stress and autocatalysis reaction. Besides, it would be quite interesting to carry out calculations and predict threshold density of electron beam energy for initiation of RDX ($C_3H_6N_6O_6$), HMX ($C_4H_8N_8O_8$) and TATB ($C_6H_6N_6O_6$).

5.2   PROBLEM STATEMENT

We analyze the thermal model of explosive initiation by the nanosecond electron beam taking into account the transformation of the explosive due to the first-order reaction. In this case, taking into consideration fusion, one-dimensional thermal conductivity equation and the kinetic equation of first-order chemical reaction are set as

$$\rho[c + H_f \, \delta(T - T_f)]\frac{\partial T}{\partial t} = \lambda\frac{\partial^2 T}{\partial x^2} + \frac{\Lambda(x)}{R_{ef}}I(t) + \rho Q\frac{d\eta}{dt} \quad , \tag{1}$$

$$\frac{d\eta}{dt} = k_1(1 - \eta) \tag{2}$$

The initial and boundary conditions of Eqs. (1) and (2) are

$$T(x,0) = T_0, \quad \eta(x,0) = 0, \quad -\lambda\frac{\partial T(0,t)}{\partial x} = 0 \quad -\lambda\frac{\partial T(h,t)}{\partial x} = 0, \tag{3}$$

where: T_0 is the initial temperature of the sample; T is temperature; T_f is fusion temperature; λ, c, ρ is the thermal conductivity coefficient, thermal capacity and density of the sample; H_f is the specific heat of fusion; Q is the thermal effect of the reaction per one mass unit of substance; R_{ef} is the effective track length of electrons; $I(t)$ is the energy flux density of the electron beam; $\Lambda(x)$ is the distribution of absorbed energy density along the crystal; h is the thickness of the sample; η is the conversion level of the explosive. The dependence of thermal and physical parameters on temperature is ignored.

First-order reaction rate constant is determined as

$$k_1 = Z\exp(-E / RT),$$

where R is gas constant; Z is reaction frequency factor; E is thermal decomposition activation energy.

The dependence of electron beam intensity on time is given by

$$I(t) = j(0,t)U_0(t) = \frac{W}{6\tau_m}(4t / \tau_m)^4 \exp(-4t / \tau_m),$$

where $j(0,t)$ is current density of electron at the input into the solid; $U_0(t)$ is accelerating stress of electron beam generator; τ_m is pulse rise time, connected with the half-amplitude duration, determined by $\tau_i = 1{,}19\tau_m$; W is the density of absorbed energy. Though the integral

$$\int_0^\infty I(t)dt = W$$

The experimental curve of density distribution of energy absorbed by the solid $W_{ab}(\xi)$ is usually approximated by a polynomial of the third order [8, 9]. In case of PETN, experimental curve of density distribution of absorbed energy [2] is satisfactorily described by equation [6, 7]:

$$W_{ab}(\xi)/W_{ab}(\xi_m) = \Lambda(\xi) = 0,7 + 1,57\xi - 2,31\xi^2 + 0,61\xi^3, \qquad (4)$$

where $\xi = x/R_{ef}$. If $\xi = \xi_m$, function $\Lambda(\xi_m) = 1$, derivative $d\Lambda(\xi_m)/d\xi = 0$. If $\xi \geq \xi_{ex} = 1,44$, function $\Lambda(\xi) = 0$. Here, $R_{ef} = 173,6 \cdot 10^{-4}$ cm is the effective track range of electrons; $R_{ex} = 1,44R_{ef} = 250 \cdot 10^{-4}$ cm is the extrapolated track range of electrons at $E_0 = 250$ keV. The distribution of absorbed energy in RDX, HMX and TATB is considered to be similar to Eq. (4).

For RDX, HMX and TATB, the extrapolated track range of electrons, which initial energy is $E_0 = 250$ keV, is calculated by the empirical formula, introduced in article [10]:

$$R_{ex} = \frac{a_1}{\rho}\left[\frac{1}{a_2}\ln(1 + a_2 e_0) - \frac{a_3 e_0}{1 + a_4 e_0^{a_5}}\right], \qquad (5)$$

Here, ρ is the density of the medium measured in g/cm³; $e_0 = E_0/mc^2$ (mc^2 is the rest energy of an electron, equal to 511 keV); constants

$$a_1 = 0,2335 A_{ef}/Z_{ef}^{1,209}, \; a_2 = 1,78 \cdot 10^{-4} Z_{ef}, \; a_3 = 0,989 - 3 \cdot 10^{-4} Z_{ef}$$

$$a_4 = 1,468 - 1,18 \cdot 10^{-2} Z_{ef}, \; a_5 = 1,232/Z_{ef}^{1,209}$$

where A_{ef} is effective atomic mass; Z_{ef} is effective atomic number. Effective atomic mass and effective atomic number are determined by formulas

$$Z_{ef} = \sum_{i=1}^{n} f_i Z_i, \; A_{ef} = \frac{Z_{ef}}{(Z/A)_{ef}}, \; (Z/A)_{ef} = \sum_{i=1}^{n} (f_i Z_i/A_i), \; f_i = A_i/\sum_{i=1}^{n} A_i,$$

Here, Z_i, A_i are atomic number and atomic mass of i-element; f_i is its weight fraction; n is the number of weight fractions. The estimation of R_{ef} for PETN, using Eq. (5) at $\xi_{ex} = 1,44$, shows that calculated value is 14 % less than R_{ef}, obtained while processing experiment results. That's why the calculated values of R_{ef} for modeling RDX, HMX and TATB initiation have been increased by 14%. The calculated values of Z_{ef}, A_{ef} and R_{ef} at $E_0 = 250$ keV are given in Table 5.1. Besides, Table 5.1 represents kinetic and thermo physical parameters of explosives.

TABLE 5.1   Kinetic and Thermo Physical Parameters of Explosives

EM	PETN	RDX	HMX	TATB
Z_{ef}	69,2	39,7	53	41,72
A_{ef}	135	77,37	103,2	81,55
R_{ef}, m	173.59×10^{-6}	240.0×10^{-6}	199.2×10^{-6}	233.7×10^{-6}
E, kJ/mol	196,6 [11]	197,3 [11]	220,8 [11]	250,9 [11]
Z	6.3×10^{19} [11]	2.02×10^{18} [11]	5×10^{19} [11]	3.18×10^{19} [11]
Q, MJ/kg	1,26 [11]	2,1 [11]	2,1 [11]	2,51 [11]
c, J/(kg·K)	1255,2 [12]	2092 [13]	1250 [14]	1000,0 [15]
λ, W/(m·K)	0,2508 [11]	0,105 [11]	0,293 [11]	0,418 [11]
ρ, kg/m^3	1.77×10^3 [15]	1.82×10^3 [16]	1.9×10^3 [16]	1.93×10^3 [19]
ΔH_f, kJ/kg	193 [18]	235,5 [18]	192.46 [19]	there are no data
T_f, K	413 [11]	476 [11]	558 [11]	>623 [17]

5.3   RESULTS AND DISCUSSION

The system of equations (the Eqs. (1) and (2)) with initial and boundary conditions (the Eq. (3)) is numerically solved using implicit difference schemes. The system of difference equations for thermal conductivity equation is solved using sweep method. Fusion accounting algorithm for the solution of thermal conductivity equation is described in Refs. [6, 20, 21]. The calculation are carried out at $h = 1$ mm. Some of the results are shown in Figs. 5.1(a–c)–5.2(a–c).

Figure 5.1a shows the calculation results of temperature distribution in PETN in the area of electron beam absorption, if energy density $W = 15.5$ J/cm^2 and pulse duration $\tau_i = 15$ ns. "Shelf" occurs on curves (1)–(3) due to the fusion of the explosive. To the right of the shelf, PETN is not fused, but to the left of it the explosive is fused. Figure 5.1b represents the calculation results of distribution of PETN transformation degree in the area of electron beam absorption. According to the figure, by the time of initiation the explosive transformation degree is negligible. The curves of temperature distribution and transformation degree along the sample are similar for HMX and RDX. While solving thermal conductivity equation for TATB, fusion is ignored as its fusion heat isn't known.

Figure 5.1c shows the calculation results of delay time of PETN, HMX and RDX initiation by the pulsed beam. The delay time is determined taking into consideration that

$$d\Delta T / dt\big|_{t=t^*} \to \infty$$

TATB has the longest delay time, and PETN has the shortest one. The curve for TABT is not shown in Fig. 5.1 c. The calculated value of delay time t^* for PETN exceeds the experimental one by 2.5 factors of ten. The critical density of electron beam energy is in good agreement with experiment [2]. Thus, "classical thermal explosion" does not allow complete describing of the experiments on PETN initiation by the electron beam. The experiment is well explained if thermo elastic stress, which is known to affect activation energy of chemical reaction [6], and radiation-thermal mechanism of initiation [7] are taken into account. It should be noted that the exothermic reaction makes major contribution to the process of explosive initiation by the nanosecond electron beam.

Figures 5.2(a)–5.2(c) indicate the calculation results of temperature dependence $\Delta T_m = T_m - T_0$ on time at the maximum of electron beam absorption near the initiation threshold in PETN, RDX and HMX. "Stairs" on the curves occur because of explosive fusion.

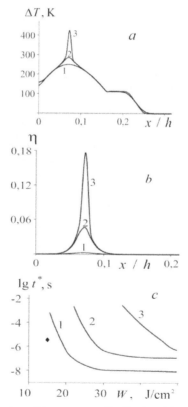

FIGURE 5.1 The dynamics of temperature distribution (*a*), transformation degrees (b) in the area of electron beam absorption while PETN initiation ($t = 10^{-4}$ (1), 6×10^{-4} (2) and 6.42×10^{-4} s (3)) and the dependence of initiation delay time for PETN (1), HMX (2) and RDX (3) on energy density of electron pulse (*c*): lines – calculations, point – the experiment for PETN [2].

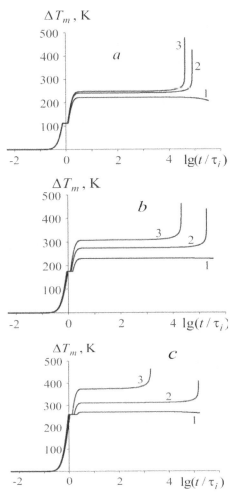

FIGURE 5.2 The dependence of temperature on time at the maximum of electron beam absorption in PETN (a), if W = 14.5 (1), 15.25 (2) and 15.5 J/cm² (3); in RDX (b) at W = 33 (1), 35 (2) and 38 J/cm² (3); in HMX (c) at W = 20 (1), 22 (2) and 25 J/cm² (3) (τ_i = 15).

The criterion of initiation by the short electron pulse for condensed energy material is obtained in article [22]. The critical initiation temperature doesn't depend on fusion and is determined by

$$\rho QZ \exp\left(-\frac{E}{RT_m}\right) = -\frac{\lambda \Delta T_m}{R_{ef}^2}\frac{\partial^2 \Lambda_m}{\partial \xi^2} \qquad (6)$$

Only the delay time of explosive initiation and critical energy density of explosive initiation by the electron beam depend on the fusion. The critical energy density of explosive initiation by the electron beam, taking into account fusion, is

$$W^* = \rho R_{ef}(c\Delta T_m + H_f) \tag{7}$$

Table 5.2 contains both the numerical solution results of systems of equations (the Eqs. (1) and (2)), and the calculation results of initiation criterion (Eqs. (6) and (7)). According to the Table 5.2, the results of numerical modeling are in a good agreement with critical energy density of explosive initiation by the short electron beam. Table 5.2 indicates that the most thermo stable explosive is TATB and also proves that the fusion affects critical energy density w^* of explosive initiation.

TABLE 5.2 Critical Temperatures of Explosive Initiation and Energy Density of Electron Beam

EM	PETN	RDX	HMX	TATB
T_m, K	520	549	593	682.3
Without fusion, J/cm²	8.48	22.7	13.9	16.44
With fusion, W^*, J/cm²	14.38	32.99	21.18	–
Numerical simulation, W^*, J/cm²	15.0	34.0	22.0	17.0

Figure 5.3 a represents calculation results of the dependence R_{ef} in PETN (1), HMX (2) and RDX (3) on initial electron energy E_0. If the energy is increased by 2.4 times, then R_{ef} increases by 3.5 times. It causes the increase in critical energy of explosive initiation in the area of electron beam absorption (Fig. 5.3b). The higher R_{ef}, the more energy is required to heat the explosive up to the critical temperature in the area of absorption. It explains the increase in W^*.

Figure 5.3c shows the values of critical energy density calculated using Eqs. (6) and (7) depending on initial electron energy in the pulse. According to Fig. 5.3c, experimental value of density energy while PETN initiation by the electron beam at E_0 = 250 keV is in good agreement with experiment [2]. However, as it's already been mentioned, the calculated delay time exceeds the experimental one by 2.5 factors of 10. Partly, it can be explained by the fact that the experiments on explosive initiation by electron or laser pulses are usually carried out a bit higher than the initiation threshold. Critical energy density W_D (E_0 = 450 keV), when explosive detonates, is two times higher than the calculated value. All in all the calculated results satisfactorily correspond to the experimental results. At least, in case of PETN initiation there is no problem in explaining the experiment results unlike the initiation of lead aside by the electron pulse [22], where calculated value of W^* exceeds the experimental one by 5–100 times depending on the fixation of the sample in the output of the pulsed accelerator.

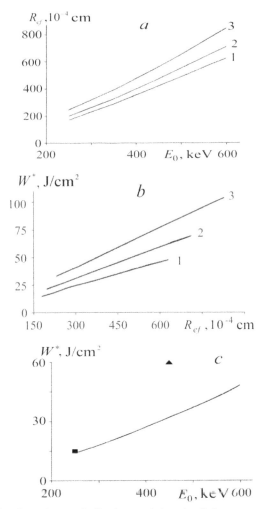

FIGURE 5.3 The dependence of effective track length of electrons on the initial energy of electrons (a), the dependence of critical energy density of the explosive initiation on R_{ef} (b) (1 – PETN, 2 – HMX, 3 – RDX) and the dependence of critical energy density of PETN initiation on initial electron energy in the beam (c): line-calculation on criterion (Eqs. (6) and (7)); ■ – experiment [2]; ▲ – the threshold energy of PETN detonation [3].

We estimate the pressure of plasma, formed in the area of energy release while the detonation of PETN, using the formula given in paper [23]:

$$P = (\gamma_{ef} - 1)w, \qquad (8)$$

where $\gamma_{ef} = 1,2$ is the ration of specific heat capacity of the solid and plasma; w is the volume density of absorbed energy. The volume density of absorbed energy for the explosive is determined by formula

$$w = \rho(Q - \Delta H_f) + W_D / R_{ef} .$$

If W_D = 60 J/cm^2, R_{ef}= 415.0×10^{-4} cm and E_0 = 450 keV, we have P=0.86 GPa. The obtained value of pressure in detonation wave for PETN is quite reasonable. According to Ref. [24], P~1.5 GPa. Maybe, PETN doesn't detonate at W = 15 J/cm^2 due to the gas-dynamic relieving of the sample, because in this case energy is released near the surface (R_{ef} = 174×10^{-4} cm, E_0 = 250 keV).

5.4   CONCLUSION

Thus, preliminary calculations and estimations make in favor of the thermal mechanism of PETN initiation by the nanosecond electron beam.

KEYWORDS

- **Electron pulse**
- **Fusion**
- **HMX**
- **Ignition criterion**
- **Initiation**
- **PETN**
- **RDX**
- **Simulation**
- **TATB**

REFERENCES

1. Korepanov, V. I., Lisitsyn, V. M., Oleshko, V. I., & Tsipilev, V. P. (2003). Detonation of PETN Single Crystals initiated by a Powerful Electron Beam, Pic'ma v JTF, *29(16)*, 23–28.
2. Aduev, B. P., Belokurov, G. M., Grechin, S. S., &, Shvaiko, V. N. (2007). Investigation of the Early Stages of Explosive Decomposition of PETN Crystals Initiated by Pulsed Electron Beams, Izv., Vyssh., Uchebn Zaved, Fiz., *50(2)*, 3–9.
3. Oleshko, V. I., Korepanov, V. I., Lisitsyn, V. M., Skripin, A. S., & Tsipilev, V. P. (2012). Electric Breakdown and Explosive Decomposition of PETN Mono crystals Initiated by an Electron Beam, Pic'ma v JTF, *38(9)*, 37–43.

4. Aduev, B. P., Belokurov, G. M., Grechin, S. S., & Puzynin, A. V. (2010). Investigation of Characteristics of Explosive Decomposition of PETN Mono crystals Initiated by Pulsed Electron Beams, Fundamental'nye Problemy Sovremennogo Materialovedenia, *2*, 43–47.

5. Oleshko, V. I., Tsipilev, V. P., Lysyk, V. V., Razin, A. V., Zarko, V. E., & Kalmykov, P. I. (2012). Explosive Decomposition of FTDO Initiated by Laser Radiation Pulse and Pulse of Accelerated Electrons, Izv, Vyssh, Uchebn, Zaved, Fiz, *55(11/3)*, 158–161.

6. Khaneft, A. V., Duginov, E. V., & Ivanov, G. A. (2012). Modeling's of PETN Initiation by an Electron Beam with Nanosecond Duration, Khimicheskay Fizika I Mezoskopiya, *14(1)*, 28–39.

7. Khaneft, A. V., & Ivanov, G. A. (2012). Radiation-Thermal Mechanism of PETN Initiation in the Absorption Region of the Electron Beam, Izv, Vyssh, Uchebn, Zaved, Fiz, *55(11/3)*, 71–75.

8. Polycrystalline Semiconductors, Physical Properties and Applications, Editor by, Harbeke, G. (1985). Springer-Verlag, Berlin, Heidelberg, New York, Tokyo, 344pp.

9. Shiller, Z., Gayzich, & Pancer (1980). Electronno-Luchevaja Tecnologija, Moscow, Energija, 528pp.

10. TatsuoTabata, & RinsukeIto (1974). An Algorithm for the Energy Deposition by Fast Electrons, Nuclear Science and Engineering, *53*, 226–239.

11. Explosion Physics, Orlenko, L. P. Ed Moscow, Nauka, 824pp.

12. Belyaev, A. V., Bobolev, V. K., Korotkov, A. I., Sulimov, A. A., & Chuiko, S. V. (1973). Transitions of Condensed Systems from Ignition to Explosions, Moscow Nauka, 292pp

13. Mader Charles, L. (1985). Numerical Modeling of Detonations, Moscow, Mir., 384pp.

14. Assovskiy Igor, G. (2005). Physics of Combustions and Interior Ballistics, Moscow, Nauka, 358pp.

15. Detonation and Explosives, Collection of Articles, Editor by, Borisov, A. A. (1981). Moscow, Mir, 392pp.

16. Shal, R. (1974). Physics of a Detonation, In the Book Physics of High Energy Density, Moscow, Mir., 488pp.

17. Garmasheva, N. V., Filin, V. P., Chemagina, I. V., Taibinov, N. P., Timofeev, V. T., Filippova, Yu N., Kazakova, M. B., Batalova, I. A., Shakhtorin, Yu A., & Loboiko, B. G. (2003). Some Features of TATB Decomposition When Heated, VII Zababakhinskie Nauchnye Chteniya, Snezhinsk, 1–13.

18. Baum, F. A., Derzhavec, A. S., & Sanasaryan, N. N. (1969). Thermoresistant Explosives and their Effect in Deep wells, Moscow, Nedra, 160pp.

19. Strunin, V. A., Nikolaeva, L. I., & Manelis, G. B. (2010). Modeling of HMX Combustion, Khimicheskaya fizika, *29(7)*, 63–70.

20. Dolgachev, V. A., & Khaneft, A. V. (2012). Simulation of PETN Initiation by a Laser Pulses of Nanosecond Duration, Khim, Fiz, Mezockop, *14(4)*, 1–8.

21. Khaneft, A. V., & Duginov, E. V. (2012). Effect of Melting on the Critical Ignition Energy of Condensed Explosives by a Short Laser Pulse, Combustion, Explosion, and Shock Waves, *48(6)*, 699–704.

22. Khaneft, A. V. (1998). Criteria for Ignition of Condensed Materials by an Electron Pulse, Chem. Phys. Reports, *8*, 1573–1581.

23. Batsanov, S. S., Demidov, B. A., & Rudakov, L. I. (1979). Use of a High-Current Relativistic Electron Beam for Structural and Chemical Transformations, JETP Letters, *30*, 575–577.

24. Morozov, V. A., Petrov, Yu V., & Savenkov, G. G. (2012). Criterion of Shock-Wave Initiation of Detonation in Solid Explosives, Dokl, Ross, Akad, Nauk, *445*, 286–288

CHAPTER 6

COMPUTER SIMULATING OF SOLID SOLUTION $Ag_xPb_{1-x}S_{1-\delta}$ HYDROCHEMICAL DEPOSITION PROCESS

A. Y. KIRSANOV, V. F. MARKOV, and L. N. MASKAEVA

CONTENTS

6.1 INTRODUCTION

Thin films of metal sulfides are widely used in various fields of technology due to the unique physical and chemical properties. Many of these are sensitive to optical radiation, physical and chemical effects. Important area of research is the creation of selective chemical sensors for in toxic gasses air monitoring [1].

Previously, to determine nitrogen oxides in the air we demonstrated efficiency of sensor elements based on thin films of lead sulfide, produced by a relatively simple hydro chemical technology. Their advantage is the high selectivity and working near room temperature, which greatly simplifies the development and structure of new devices.

Formation of substitution solid solutions based on PbS allows to expand the range of PbS electro and sensory properties. Solid solution of films AgxPb1-xS are of great interest as such materials. However, sulfides PbS and Ag_2S have very low mutual solubility under standard conditions. According to the high-temperature phase diagram of the PbS-Ag_2S for solid samples limiting solubility Ag_2S in PbS is achieved at 970 K and no more than 0.4 mol.% [3], which corresponds to the solid solution $Ag_{0.008}Pb_{0.992}S$. There are no published information about the formation of solid solutions $Ag_xPb_{1-x}S$ silver content in the structure more than 0.8 at. %. Contemplated sulfides have various types of crystal lattices and spatial groups. Moreover, contemplated sulfides have different compounds in the valent state. Lead sulfide form, respectively, a simple cubic (structure B1), and monoclinic Ag_2S or body-centered crystal lattice [3]. Hydro chemical deposition methods allow extending the range of compositions of substitutional solid solutions $Ag_xPb_{1-x}S$, despite the poor conditions of isomorphic substitution [4, 5]. This method based on colloid-chemical step; provide conditions for the creation of Meta stable supersaturated compounds. Computer simulation is a proper way to solving the problem of managed hydro chemical synthesis by co precipitation PbS and Ag_2S. Computer simulating allows predicting the composition of the solid solutions $Cd_xPb_{1-x}S$ [6] and $Cu_xPb_{1-x}S$ [7] in their hydro chemical synthesis showed a relatively high convergence of the results with experimental data.

The first purpose of this study was to develop a computer model of the process of co precipitating hydro chemical deposition of lead and silver sulfides to select potential synthesis conditions substitutional solid solutions $Ag_xPb_{1-x}S$. The second purpose is experimental validation of the proposed model.

6.2 EXPERIMENTAL TECHNIQUE

Film formation and growth computer model $Ag_xPb_{1-x}S$ is based on process of nucleus formation and primary clusters in reaction volume and further aggregative growth [8].

VASP was taken as a home frame for calculations [9]. Requisite expansion of VASP functional has been devised by authors.

Regulations of rapid coagulation theory of Smolukhovskogho were used to set fundamental principle of solid phase formation, subject to stochastic nature of particles' motion.

To conduct calculation in acceptable time, volume of reaction mixture was conventionally divided into equal cubic "micro volumes," which is taken as pseudo-valid reactor with acceptable statistic deviations. Subsequently holistic picture of hydro chemical synthesis process was obtained by composition of descriptive status of "micro volumes." In reaction mixture balancing of reagent concentration in "micro volumes" was calculated by thermodynamic diffusion flows.

Second necessary criterion of the calculations in acceptable time is to set boundaries of concentration ranges per each of reagents. For the sake of this authors found concentration fields of individual metal sulfide conformation. For revelation of sulfide formation PbS and Ag_2S collaborative area in temperature interval 298–343K analysis of ion equilibrium in multi component system was conducted, including lead acetate $Pb(CH_3COO)_2$, silver nitrate $AgNO_3$, trisubstituted sodium citrate $Na_3C_6H_5O_7$, ammonium hydroxide NH_4OH and thiocarbamide N_2H_4CS. Presence of citrate ions and thiocarbamide, binding in durable complex lead and silver, respectively, in the reaction mixture prevents rapid emission of metal sulfide in settling.

In this research, only concentration of lead salts and silver were differed in reaction mixture composition, content of other reagents was constant in every of calculated experiment.

Concentrations, involved in reagent synthesis and its' characterizing constants, were set by parameters, which were initialized in calculation program.

An important feature of developed algorithm is simultaneous multithreaded control of all processes, coursed by synthesis in variety of reaction mixture microvolumes.

In case of calculating of electronic structure, stability of lattices and interaction energy were used following approximations of quantum chemistry, necessary for solving Schrodinger equation. Born-Oppenheimer approximation, Hartree approximation, MO LCAO approximation (molecular orbital as a linear combination of atomic orbital).

Calculations were made only on base of prior software with new source data, later were used its' average results. Quantum chemical calculations were conducted with application of "first principles" simulation program (the solution of first fundamental principles without additional empirical assumptions).

Density functional theory (DFT), which allows replacing many electron wave function by electron density, using ab initio methods with given approaches and simplifications enables to provide required calculations.

Simulating package houses major simplifications: valence approximation, approximation of local electron density (LDA+U); replacement of many-electron problem solution to the one-electron (with effective local potential); description of kinetic energy of electron movement by local approximation on the basis of free-electron theory; self-consistent field method.

For the purpose of determination of solid solution electron structure, were additionally used following approximations and calculation methods: generalized gradient approximations (GGA); self-interaction correction (SIC); optimized effective potential method; GW approximation (replacement of Coulomb potential in the Hartree-Fock approximation to the dynamically dispersive potential, furthermore exchange-correlation potential replaces by Green's function); full-potential method of linear muffin-tin orbital (FP LMTO); pseudo potential method.

6.3   EXPERIMENTAL RESULTS

The result of the simulation software is a description of the entire volume of the reactor to the level of detail for each atom, and in particular deposited onto a substrate material, which is divided into "clusters." Accordingly, the composition of the resulting solid solutions $Ag_xPb_{1-x}S$ was defined internally developed software using the average value of a sequence of atomic conversion probability sample set of "clusters." Allowable error (statistical sampling error) was 4%.

Graphical interpretation of the calculation results in the coordinates "Content Ag_2S solid solution function of the initial concentrations of the metal salts in the reaction mixture" is shown in Fig. 6.1.

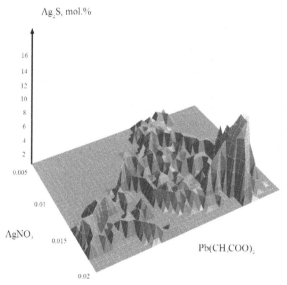

FIGURE 6.1   The graphical interpretation of the results of computer simulation: the dependence of silver sulfide content in the solid solution $Ag_xPb_{1-x}S$ from the initial concentration of metal salt in the reaction mixture.

As in the case of computer modeling of the solid solutions in the deposition system PbS-CdS [6], the surface has pronounced local maximum. They substitute the appropriate content component in the solid solution $Ag_xPb_{1-x}S$ in comparison with high phase diagram [3] greatly exceeds the limiting solubility of Ag_2S in the structure of PbS, which allows us to state of the formation of supersaturated solid solutions of substitution. Maximum content of Ag_2S in $Ag_xPb_{1-x}S$ was 12.8 mol%. However, apparently, we cannot clearly determine it as the absolute maximum. Besides solid solution $Ag_xPb_{1-x}S$, in "clusters" found a significant content of individual phases of silver sulfide, which introduces additional complexity in the interpretation of the data. It should be noted that there is a pronounced interfacial distribution of lead and silver in the film depending on the ratio of components in the reaction mixture.

We performed a series of experiments on hydro chemical deposition films $Ag_xPb_{1-x}S$ by co precipitation individual metal sulfide PbS and Ag_2S to test the adequacy of the results. Synthesis was carried out for 90 min at a temperature of 303 K during a change in the reaction mixture lead acetate content from 0.01 to 0.04 mol/L, and while varying the concentration of the silver salt in the range of 5.0×10^{-5}–1.2×10^{-2} mol/L.

Analyzing the XRD patterns obtained films was observed the structure of cubic phase B1 only, the period of which is dependent on the film composition. Fig. 2 shows the X-ray films of lead sulfide and individual co precipitating layers. We found that the period of the cubic phase of lead sulfide with increasing silver content in the reactor increased from 0.5932 to 0.5935 nm. It was interpreted as the formation of PbS by substitutional solid solutions $Ag_xPb_{1-x}S$. In the formation of solid solution the lattice period increased due to replacement of lead ions (II) with a radius of 0.120 nm lattice PbS to larger silver Ag^+ ions with a radius of 0.126 nm.

micro analyzer Superzond JCXA-733c on three points on the content of lead, silver and sulfur. We evaluated the number of elements in the films obtained by varying the composition of the reaction mixture.

XRD patterns were obtained using an X-ray spectrometer EDS Inca Energy 250. Structural studies of the films were performed on a DRON-UM1. XRD patterns was obtained by using copper radiation pyrolytic graphite monochromator to extract $CuK\alpha1$, 2-doublet of continuous spectrum in the range of angles 2θ of $20°$ to $90°$ in step-scan mode in steps of $\Delta(2\theta) = 0.02$ and the accumulation time of the signal at 5 s.

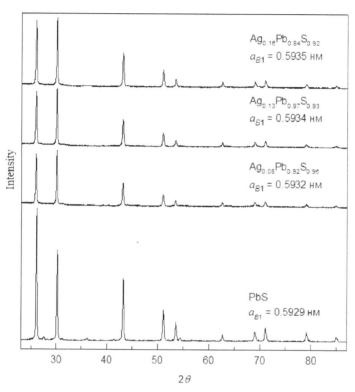

FIGURE 6.2 XRD patterns PbS films and solid solutions $Ag_{0.08}Pb_{0.92}S_{0.96}$, $Ag_{0.13}Pb_{0.87}S_{0.93}$ and $Ag_{0.16}Pb_{0.84}S_{0.92}$, obtained by co-precipitation of sulfides of lead and silver at T = 303 K.

Elemental analysis of the films co precipitated sulfide PbS and Ag_2S formed

Comparative results of calculations and experimental data are given in the (Table 6.1). The table shows that with increasing concentration of silver salt in the reaction mixture in its solid phase content increases monotonously while reducing the lead. Among the samples studied Ag_2S maximum content in the $Ag_xPb_{1-x}S$ solid solution was 8.5 mol%. The discrepancy between the results of simulation and experiment does not exceed 14.4%, which is certainly acceptable. Note worthy significant non-stoichiometry of sulfur in the composition of the solid solution. This occurrence may be explained by oxygen-containing metal phase in film, in particular, sodium hydroxide, lead, whose formation is confirmed by calculations of ionic equilibria. Detected non-stoichiometry on the sulfur content in the deposited solid solutions in general is reflected in the writing of their composition $Ag_xPb_{1-x}S_{1-\delta}$.

6.4 SUMMARY AND CONCLUSIONS

Computer simulation of the $Ag_xPb_{1-x}S_{1-\delta}$ solid solutions hydro chemical synthesis process was carried out. The potential conditions for the formation of supersaturated solid solutions with a replacement component, exceeding the solubility limit Ag_2S in PbS according to high temperature phase diagram were founded. Ag_2S concentration in solid solution has a number of local maximums on a calculated surface. Its absolutes maximum of 12.8 mol%.

The experimental results confirmed the adequacy of the simulation results. The maximum discrepancy between experiment and computer simulation results was 14.4%. Receipt of the solid solutions $Ag_xPb_{1-x}S_{1-\delta}$ with the concentration of component replacement stated above the solubility limit in the areas calculated local maximum was confirmed.

TABLE 6.1 The Experimental Data and Computer Simulations on the Effect of Synthesis Conditions on the Composition of Hydro Chemical Co-precipitated Solid Solution Films $Ag_xPb_{1-x}S$.

Synthesis conditions	Content of elements in the film, at %						Formulaic composition of the $Ag_xPb_{1-x}S_{1-d}$ solid solution (simulating result)	Formulaic composition of the $Ag_xPb_{1-x}S_{1-d}$ solid solution (based on lattice period)
	Experiment			Simulation				
[AgNO₃], mol/L	Ag ±0.7	Pb ±0.8	S ±0.5	Ag	Pb	S		
0.001	3.9±0.6	53.8±0.4	42.3±0.8	3.5	55.8	40.7	$Ag_{0.07}Pb_{0.93}S_{0.96}$	$Ag_{0.08}Pb_{0.92}S_{0.96}$ ($a = 0.59319$ нм)
0.002	4.3±0.3	51.6±0.6	44.1±0.3	4.9	50.8	44.3	$Ag_{0.11}Pb_{0.89}S_{0.95}$	$Ag_{0.10}Pb_{0.90}S_{0.95}$ ($a = 0.59324$ нм)
0.005	27.1±0.8	32.4±0.8	40.5±0.5	29.0	31.1	39.9	$Ag_{0.15}Pb_{0.85}S_{0.93}$	$Ag_{0.13}Pb_{0.87}S_{0.93}$ ($a = 0.59338$ нм)
0.010	27.4±1.0	32.3±0.9	40.3±0.5	31.1	30.4	38.5	$Ag_{0.16}Pb_{0.84}S_{0.92}$	–
0.012	33.1±0.8	30.2±0.6	36.7±0.6	35.8	29.2	35.3	$Ag_{0.17}Pb_{0.83}S_{0.92}$	$Ag_{0.16}Pb_{0.84}S_{0.92}$ ($a = 0.59348$ нм)

ACKNOWLEDGMENTS

This work was supported by RFBR grant 13-03-96093.

KEYWORDS

- **Computer simulating**
- **Hydrochemical deposition**
- **Lead sulfide**
- **Silver sulfide**
- **Solid solution**

REFERENCES

1. Vasiliev, R. B., Ryabova, L. I., Rumyantseva, M. N., & Gaskov, A. M. (2004). In organic Structures as Materials for Gas Sensors, Russian Chemical Reviews.
2. Markov, V. F., & Maskaeva, L. N., (2001). The Semiconductor Sensor of Nitrogen Oxides Analyzer Based Lead Sulfide, J. Anal. Chem., 73.
3. Shelimova, L. E., Tomashik, V. N., & Grytsiv, V. I. (1991). State Diagrams in Semiconductor Materials (Chalcogenide Based Systems Si, Ge, Sn, Pb) (Science, Moscow).
4. Markov, V. F., Maskaeva, L. N., & Ivanov, P. N. (2006). Hydro Chemical Film Deposition of metal Sulfides, Simulating and Experiment (UB RAS, Ekaterinburg).
5. Maskaeva, L. N., Markov, V. F., & Moskaleva, A. A. (2011). Hydro Chemical Synthesis of Metal Chalcogenides Films, Part 7 Preparation of Thin Films of PbS-Ag$_2$S by Ion Exchange Replacement, Butlerov Communications, 26.
6. Kirsanov, A. Y., Markov, V. F., & Maskaeva, L. N. (2013). The Simulation of the Formation of Solid solutions Cd$_x$Pb$_{1-x}$S Hydro Chemical Deposition, Bulletin of the South Ural State University, 5.
7. Kirsanov, A. Y., Markov, V. F., & Maskaeva, L. N. (2012). The Simulation of the Hydro chemical Synthesis of Super saturated Solid Solutions in the System PbS-CuS, Fundamental Problems of Modern Materials, 9.
8. Markov, V. F., & Maskaeva, L. N. (2011). Features of Formation of Metal Sulfide Film from Water Solutions, Butler ov Communications, 24
9. http://www.vasp.at.

CHAPTER 7

CALCULATIONS OF THE AFFINITY FOR ELECTRON VIA THE VALUES OF SPATIAL-ENERGY PARAMETERS

G. A. KORABLEV and G. E. ZAIKOV

CONTENTS

7.1 INTRODUCTION

The amount of energy isolated when an electron connects to a neutral atom in its main state is called the affinity for electron (χ). That is such energy equals the difference of the energy of atom (molecule) in the main state and energy of the main state of the corresponding negative ion.

A lot of investigations are dedicated to the experimental and theoretical definition of this value [1, 2, 3]. The analysis of this data allows making the conclusion that there are often big discrepancies in the values obtained (quite frequently even in the framework of one method). Thus, for iron atom we obtained χ-equaled to 0.58 eV and 1.06 eV, for radical C_2H: 2.1 eV; 2.7 eV and 3.73 eV, for BO: 2.12 eV and 3.1 eV etc. There are no doubts about the reliability of the techniques. Obviously in these cases different structural modifications of atoms and radicals are considered. Therefore, it is interesting to find the dependence of the energy of affinity for electron with initial energy characteristics of atom in its different valence states. For this the concept of spatial-energy parameter (P-parameter) is used in this investigation.

7.2 SPATIAL-ENERGY P-PARAMETER

The analysis of the kinetics of various physic-chemical processes demonstrates that in many cases the reciprocals of velocities, kinetic or energy characteristics of the corresponding interactions are added.

Here are some examples: ambipolar diffusion, total rate of topochemical reaction, changes in the light velocity when transiting from vacuum into the given medium, effective permeability of bio-membranes.

In particular, such assumption is confirmed by the formula of electron transport probability (W_∞) due to the overlapping of wave functions 1 and 2 (in stationary state) during electron-conformation interactions:

$$W_\infty = \frac{1}{2}\frac{W_1 W_2}{W_1 + W_2} \tag{1}$$

The Eq. (1) is applied [4] when evaluating the characteristics of diffusion processes accompanied with non-radiating electron transport in proteins.

Lagrangian equation is also illustrative. For the relative motion of the isolated system of two interacting material points with masses m_1 and m_2 in coordinate x it looks as follows:

$$m_r x'' = -\frac{\partial U}{\partial x}, \text{ where } \frac{1}{m_r} = \frac{1}{m_1} + \frac{1}{m_2}$$

where U – mutual potential energy of points; m_r – reduced mass. At the same time $x'' = a$ (characteristic of system acceleration). For elementary regions of interactions Δx can be taken as follows:

$$\frac{\partial U}{\partial x} \approx \frac{\Delta U}{\Delta x}. \text{ That is: } m_r a\Delta x = -\Delta U.$$

Then:

$$\frac{1}{1(\ a\Delta x)\ (1/m_1 + 1/m_2)} \approx -\Delta U;$$

$$\frac{1}{1(\ m_1 a\Delta x) + 1(\ m_2 a\Delta x)} \approx -\Delta U$$

Or: $$\frac{1}{\Delta U} \approx \frac{1}{\Delta U_1} + \frac{1}{\Delta U_2} \qquad (2)$$

where ΔU_1 and ΔU_2 – potential energies of material points on the elementary region of interactions, ΔU – resulting (mutual) potential energy of these interactions.

"Electron with the mass m moving near the proton with the mass M is equivalent to the particle with the Mass $m_r = \dfrac{mM}{m+M}$ " [5].

In this system the energy characteristics of subsystems are: orbital energy of electrons (W_i) and nucleus effective energy taking screening effects into account (by Clementi [10]).

Therefore, assuming that the resulting energy of "orbital-nucleus" system interaction (responsible for interatomic interactions) can be calculated adding the reciprocals of some initial energy components, the introduction of P-parameter as an averaged energy characteristic of valence orbitals was substantiated [6] based on the following equations:

$$\frac{1}{q^2/r_i} + \frac{1}{W_i n_i} = \frac{1}{P_y} \qquad (3)$$

$$P_э = \frac{P_0}{r_i} \qquad (4)$$

$$\frac{1}{P_0} = \frac{1}{q^2} + \frac{1}{(Wrn)_i} \qquad (5)$$

$$q = \frac{Z*}{n*} \qquad (6)$$

where W_i – bond energy of electrons [7]; or atom ionization energy (E_i) [8]; r_i – orbital radius of i orbital [9]; n_i – number of electrons of the given orbital, Z^* and n^* – nucleus effective charge and effective main quantum number [10, 11].

P_0 is called a spatial-energy parameter, and P_E – an effective P-parameter [12]. The effective P_E-parameter has a physical sense of some averaged energy of valence electrons in an atom and is measured in energy units, for example, electron-volts (eV).

Based on the Eqs. (3–6) values of P-parameters have been calculated, some results are given in Table 7.1.

Let us briefly explain the reliability of such an approach. As the calculations demonstrated [6] the values of P_E-parameters equal numerically (in the range of 2%) the total energy of valence electrons (U) by the atom statistic model. Using the known correlation between the electron density (b) and intra-atomic potential by the atom statistic model, we can obtain the direct dependence of P_E-parameter on the electron density at the distance r_i from the nucleus.

The rationality of this equation was proved by the calculation of electron density using wave functions by Clementi [13] and comparing it with the value of electron density calculated through the value of P_E-parameter.

Based on the calculations and comparisons two principles of adding spatial-energy criteria depending on wave properties and systemic character of interactions and particle charges are substantiated:

1. Interactions of oppositely charged (heterogeneous) systems consisting of I, II, III, .. atom grades are satisfactorily described by the principle of summing up their reciprocals by the Eqs. (3–6).

2. When similarly charged (homogeneous) subsystems are interacting, the principle of algebraic addition of their P-parameters is fulfilled by the following equations:

$$\sum_{i=1}^{m} P_0 = P_0' + P_0'' + \ldots + P_0^m \tag{7}$$

$$\sum P_E = \frac{\sum P_0}{R} \tag{8}$$

where R is the dimensional characteristic of tom (or chemical bond).

7.2.1 WAVE EQUATION OF P-PARAMETER

To characterize spatial-energy properties of an atom two types of P-parameters are introduced with the simple link between them:

TABLE 7.1 P-Parameters of Atoms Calculated Via the Bond Energy of Electrons

Atom	Valence electrons	W (eV)	r_i (Å)	q^2_0 (eV Å)	P_0 (eV Å)	R (Å)	P_0/R (eV)	$\dfrac{P_0}{R(n^*+1)}$	r_i (Å)	P_0/r_i (eV)	$\dfrac{P_0}{r_i(n^*+1)}$
H	1S¹	13.595	0.5295	14.394	4.7985	0.5295 0.28	9.0624 17.137	4.5312 8.5685	1.36	3.525	1.7625
Li	2S¹	5.3416	1.586	5.86902	3.475	1.55	2.2419	0.7473	0.68	5.1103	1.7034
Be	2S¹	6.4157–	1.040	13.159–	5.256–	1.13–	4.6513–	1.5504–	0.34–	15.459–	5.153–
	2S²	8.4157	1.040	13.159	7.512	1.13	6.6478	2.2159	0.34	22.094	7.3647
B	2P¹	8.4315–	0.776–	21.105–	4.9945–	0.91–	5.4885–	1.8295–	0.20	24.973	8.3242
	2S¹	13.462–	0.769–	23.890–	6.9013–	0.91–	7.5838–	2.5279–			
	2S²	13.462	0.769	23.890	11.092–	0.91–	12.189–	4.0630–	0.20	80.43	26.81
	2P¹+2S²				16.086	0.91	17.677	5.892			
C	2P¹	11.792–	0.596–	35.395–	5.6680–	0.77–	7.6208–	2.5403–	2.60–	2.2569–	0.75231–
	2P²	11.792–	0.596–	35.395–	10.061–	0.77–	13.066–	4.3554–	4.3554–	3.8696–	1.2899
	2S¹	19.201–	0.620–	37.240–	9.0209–	0.77–	11.715–	3.9052–	2.60–	3.4696–	1.1565–
	2S²	19.201	0.620	37.240	14.524–	0.77–	18.862–	6.2874–	2.60–	5.5862–	1.8621–
	2P²+2P²				24.585	0.77	31.929	10.643	2.60	9.4558	3.1519
N	2P¹	15.445–	0.4875–	52.912–	6.5916–	0.71–	9.2839–	3.0946–	1.48	10.696	3.5653
	2P³	15.445–	0.4875–	52.912–	15.830–	0.71	22.296–	7.4319			
	2S¹	25.724–	0.521–	53.283–	40.709–		15.083–				
	2S²	25.724	0.521	53.283	17.833–	0.71	25.117–	15.804	1.48	22.745	7.5818
	2S²+2P³				33.663		47.413				

$$P_E = \frac{P_0}{R}$$

Taking into account additional quantum characteristics of atom sublevels, this equation can be written down in coordinate x as follows:

$$\Delta P_E = \frac{\Delta P_0}{\Delta x} \text{ or } \partial P_E = \frac{\partial P_0}{\partial x},$$

where the value ΔP equals the difference between the P_0-parameter of i orbital and P_{CD}-parameter of countdown (parameter of the main state at the given set of quantum numbers).

According to the established [6] rule of adding P-parameters of similarly charged or homogeneous systems for two orbitals in the given atom with different quantum characteristics and in accordance with the law of energy conservation we have:

$$\Delta P_E'' - \Delta P_E' = P_{E,\lambda},$$

where $P_{E,\lambda}$ is the spatial-energy parameter of quantum transition.

Having taken $\Delta\lambda = \Delta x$ as the dimensional characteristic, we have:

$$\frac{\Delta P_0''}{\Delta\lambda} - \frac{\Delta P_0'}{\Delta\lambda} = \frac{P_0}{\Delta\lambda} \text{ or: } \frac{\Delta P_0'}{\Delta\lambda} - \frac{\Delta P_0''}{\Delta\lambda} = -\frac{P_0}{\Delta\lambda}$$

Let us divide once again term by terms by $\Delta\lambda$: $\left(\frac{\Delta P_0'}{\Delta\lambda} - \frac{\Delta P_0''}{\Delta\lambda}\right)\bigg/ \Delta\lambda = -\frac{P_0}{\Delta\lambda^2},$

where: $\left(\frac{\Delta P_0'}{\Delta\lambda} - \frac{\Delta P_0''}{\Delta\lambda}\right)\bigg/ \Delta\lambda \approx -\frac{d^2 P_0}{d\lambda^2}$, i.e.: $\frac{d^2 P_0}{d\lambda^2} + \frac{P_0}{\Delta\lambda^2} \approx 0$

Taking into account only the interactions when $2\pi\Delta x = \Delta\lambda$ (closed oscillator), we have the following equation:

$$\frac{d^2 P_0}{dx^2} + 4\pi^2 \frac{P_0}{\Delta\lambda^2} \approx 0$$

As $\Delta\lambda = \frac{h}{mv}$, then: $\frac{d^2 P_0}{dx^2} + 4\pi^2 \frac{P_0}{\Delta h^2} m^2 v^2 \approx 0$

or
$$\frac{d^2 P_0}{dx^2} + \frac{8\pi^2 m}{h^2} P_0 E_k = 0 \qquad (9)$$

where $E_k = \dfrac{mv^2}{2}$ is the electron kinetic energy.

Schrodinger equation for the stationary state in coordinate x:

$$\frac{d^2 \psi}{dx^2} + \frac{8\pi^2 m}{h^2} \psi E_k = 0 \qquad (10)$$

When comparing the Eqs. (9) and (10) we see that P_0-parameter correlates numerically with the value of Ψ-function: $P_0 = \Psi$.

And in general case it is proportional to it: $P_0 \sim \Psi$. Taking into consideration the wide practical opportunities of applying the methodology of P-parameter, we can take this criterion as the materialized analog of Ψ-function.

Since P_0-parameters and Ψ-function have wave properties, the principles of superposition should be fulfilled for them, which determine the linear character of the equations of adding and amending P-parameters.

7.2.2 WAVE PROPERTIES OF P-PARAMETERS AND PRINCIPLES OF THEIR ADDITION

Since P-parameter has wave properties (similar to Y'-function), the regularities of the interference of the corresponding waves should be mainly fulfilled at structural interactions.

The interference minimum, weakening of oscillations (in antiphase) occurs if the difference of wave move (Δ) equals the odd number of semi-waves:

$$\Delta = (2n+1)\frac{\lambda}{2} = \lambda\left(n + \frac{1}{2}\right), \text{ where n = 0, 1, 2, 3,} \qquad (11)$$

As applicable to P-parameters this rule means that the interaction minimum occurs if P-parameters of interacting structures are also "in antiphase" either oppositely charged or heterogeneous atoms (for example, during the formation of valence-active radicals CH, CH_2, CH_3, NO_2 ..., etc.) are interacting.

In this case P-parameters are summed by the principle of adding reciprocals of P-parameters Eqs. (3) and (5).

The difference of wave move (Δ) for P-parameters can be evaluated via their relative value $\left(\gamma = \dfrac{P}{P_i}\right)$ or the relative difference of P-parameters (coefficienta)

which at the interaction minimum produce an odd number:

$$\left(\gamma = \frac{P}{P_i}\right) = \left(n + \frac{1}{2}\right) = \frac{3}{2}; \frac{5}{2} \dots \text{ When } n = 0 \text{ (main state) } \frac{P}{P_i} = \frac{1}{2}.$$

It should be pointed out that for stationary levels of one-dimensional harmonic oscillator the energy of these levels $e = hn(n + \frac{1}{2})$, therefore in quantum oscillator, in contrast to the classical one, the least possible energy value does not equal zero.

In this model the interaction minimum does not provide zero energy corresponding to the principle of adding reciprocals of P-parameters Eqs. (3) and (5).

The interference maximum, strengthening of oscillations (in phase) occurs if the difference of wave move equals the even number of semi-waves:

$$\Delta = 2n\frac{\lambda}{2} = \lambda n \text{ and } \Delta = \lambda(n + 1).$$

As applicable to P-parameters the maximum interaction intensification in the phase corresponds to the interactions of similarly charged systems or systems homogeneous by their properties and functions (for example, between the fragments or blocks of complex inorganic structures, such as CH_2 and NNO_2 in octogene).

Then:

$$\gamma = \frac{P}{P_i} = (n + 1) \tag{12}$$

By the analogy, for "degenerated" systems (with similar values of functions) of two-dimensional harmonic oscillator the energy of stationary states:

$$\varepsilon = h\nu(n + 1)$$

By this model the interaction maximum corresponds to the principle of algebraic addition of P-parameters – Eq. (7). When $n = 0$ (main state) we have $P = P_i$, or: the interaction maximum of structures occurs if their P-parameters are equal. This rule was used as the main condition of isomorphic for the formation of isomorphic replacement [6].

7.2.3   CALCULATIONS AND COMPARISONS

From the Eq. (12) we have: $P_i = \dfrac{P}{n+1}$ or $P_i = \dfrac{P}{n}$,

where $n = 0, 1, 2, \dots$ (whole number).

This means that apart from the initial (main) state of the atom with P-parameter, each atom can have structural-active valence orbital with another value of P_i-parameters, besides the closest most active valence states differ by the values of

P-parameters in 2 times. Formally this corresponds to the increase in the distance of interatomic (intermolecular) interaction in 2 times, that is, there is the transition from the interaction radius to diameter.

Therefore for P_E-parameter we have:

$$P_E = \frac{P_0}{R(n+1)},$$ (12a)

or

$$P_E = \frac{P_0}{R\,n}$$ (12b)

Apparently the element system periodicity also corresponds to the Eq. (12a) in which, taking into account screening effects, it is better to use the effective main quantum number n^* instead of n, the connection between which by Slater [14] is as follows:

$$n\ 1–2\ 3–4\ 5–6$$

$$n^*\ 1–2\ 3–3.7–4\ 4.2$$

Then:

$$P_E = \frac{P_i}{R(n^*+1)}$$ (13)

where $P_i–P_0$ – parameters of each element in the system given period.

TABLE 7.2 Calculations of the Affinity for Electron for Atoms (eV)

Atom	Orbital	$P_E = \dfrac{P_0}{Rk\left(n^*+1\right)}$	N	$P = \dfrac{P_E}{N}$ (eV)	χ [1,2]
1	2	3	4	5	6
Li	$2S^1$	0.7473 $P_i = 1.7034$	1 3	0.7473 0.5678	0.65–1.05 0.58
Be	$2S^1$	1.5504	4	0.3885	0.38
B	$2P^1$	1.8295	5	0.369	0.30 (0.33)
C	$2P^1$ $2P^2$	2.5403 4.3554	2 4	1.2702 1.0889	1.27 1.17
N	$2P^1$	3.0946/2	7	0.221	–0.21
O	$2P^2$	5.982/2	2	1.4955	1.407 (1.48)
F	$2P^1$	3.4557	1	3.4557	3.448

F	$2S^2 2P^3$	2.6463	7	3.7804	3.45
Na	$3S^1$	0.60892/2	1	0.3045	0.34
Na	$3S^1$	0.6089	1	0.6089	0.54 ± 0.1
Mg	$3S^1$	0.91544/2	2	0.2287	-0.22
Al	$3P^1$	1.021/2	1	0.5105	0.5 (0.52)
Si	$3P^1$	1.4259 1.245	1 1	<1.335>	(1.39) 1.36
P	$3P^1$	1.7696/2	1	0.8848	0.8
S	$3P^1$	1.9265	1	1.9265	2.077
Cl	$3P^5$	6.7989/2	1	3.3995	3.614
K	$4S^1$	0.4372	1	0.4372	0.55 0.47
Ca^{+2}	$4S^2$	1.8097	1	1.8097	-1.93
Sc	$4S^1$	0.7950	1	0.7950	-0.73
Ti	$4S^1$	0.9075	1	0.4538	0.39
V(II)	$4S^1 3d^1$	1.9447	2	0.9724	(0.94)
Cr(I) Cr(III)	$4S^1$ $4S^2 3d^1$	1.1345 2.6763	1 3	1.1345 0.8921	0.98 1.2
Mn	$4S^1$ $4S^2 3d^1$	1.0501 2.7379	1 3	1.0504 0.9126	(1.07) -0.97
Fe(I) Fe(I) Fe(II)	$4S^1$ $4S^1$ $4S^1 3d^1$	1.0991 1.099/2 2.1475	1 1 2	1.0991 0.54955 1.07375	1.06 1.06 0.58 ± 0.2
Co(I) Co(II) Co(III)	$4S^1$ $4S^1 3d^1$ $4S^2 3d^1$	1.1282 2.1629 2.8502	1 2 3	1.1282 1.0815 0.9521	1.06 1.06 0.94 ± 0.15
Ni(I) Ni(II) Ni(III)	$4S^1$ $4S^2$ $4S^2 3d^1$	1.1363 1.6557 3.8934	1 1 3	1.1363 1.6557 1.2978	 1.62 1.2802
Cu(I) Cu(II)	$4S^1$ $4S^2$	1.1795 1.9023	1 1	1.1795 1.9023	1.226 1.8 ± 0.1
Ga	$4P^1$	0.90435/2	1	0.4522	0.39
Rb	$5S^1$	0.4325	1	0.4325	0.42
Sr	$5S^1$	0.56409	1	0.56409	-0.5
Y	$5S^1$ $4d^1$	0.71276/2 0.62714/2	1 1	0.3564 0.3136	-0.4 0.3
Zr	$5S^1$	0.8166/2	1	0.4083	0.45

Zr	$5S^1$ $5S^2$	0.8166 1.2829	1	<1.0498>	1.0
Nb	$4d^1$	1.0608	1	1.0608	1.13
Nb*	$5S^2$	1.4975	1	1.4975	1.3
Mo	$4d^1$	1.1946	1	1.1946	1.18
Tc	$5S^1$	0.99721	1	0.99721	0.99 1.00
Ru*	$4d^1$	1.3501	1	1.3501	1.4
Rn	$5S^2$ $4d^1$	1.7718 1.3795	1 1	1.7718	1.68
Pd*	$5S^1$	1.0609	1	1.0609	1.02
Pd	$4d^1$	1.3956	1	1.3956	1.4
Ag Ag*	$4d^1$ $5S^2$	1.3589 1.8398	1 1	1.3589 1.8398	1.3 2.0
In	$5P^1$	0.75778	1	0.75778	0.72
Sn	$5P^1$	1.0607	1	1.0607	1.03
Sb	$5P^1$ $5S^2 5P^4$	1.1140 5.2798	1 5	1.1140 1.056	0.94 0.94
Te	$5P^2$	1.9024	1	1.9024	~2
I-I	$5P^1$	2×1.4509	1	2.9218	3.0
Cs	$6S^1$	0.3992	1	0.3992	0.39
Ba	$6S^2$	0.8685	2	0.4343	−0.48
La	$6S^2$	1.1579	2	0.57895	0.55
Hg	$6S^2$	1.3538	2	0.6769	0.63
W W	$5d^1$ $6S^1$	1.1736 0.9413/2	1 1	1.1736 0.4707	1.23 0.5±0.3
Re	$6S^1$	0.9613/2	1	0.4807	0.38
Os	$5d^1$	1.3606	1	1.3606	1.44
Ir	$6S^2$	1.7079	1	1.7079	1.97
Pt	$5d^2$ $6S^2 5d^2$	2.7428 4.4522	1 2	2.7428 2.2269	2.56 2.128
Au	$5d^2$ $6S^1 5d^1$	2.7339 2.3731	1 1	2.7339 2.3731	2.8±0.1 2.3086
Te	$6P^1 6S^2$	2.9395/2	3	0.4899	0.5±0.1
Pb	$6P^1$ $6P^2$	0.95235 1.7256	1 1	0.95235 1.7256	1.03 1.56
Po	$6P^1$	1.1990	1	1.1990	1.32

TABLE 7.3 Affinity for Electron for Binary Systems ($\mathcal{X}$) in eV

Structure	First component				Second component				P_c (eV)	$\mathcal{X}$ [1,2]
	Orbital	P_x	N	P_x/N	P_x	Orbital	N	P_x/N		
1	2	3	4	5	6	7	8	9	10	11
CH	2P¹	2.5403	1	2.5403	1S¹	4.5312	1	4.5312	1.6278	1.65
CH	2P¹	2.5403/2	1	1.27015	1S¹	4.5312/2	1	2.2656	0.8139	0.74
C₂	2P¹2S¹	6.8958	1	6.8958	2P¹2S¹	6.8958	1	6.8958	3.4479	3.3;3.54
CH₂	2P¹	2.5403	1	2.5403	2S¹	2×8.566	1	17.192	2.2123	2.3
C₂H	2P¹2S¹	2×3.4479	2	3.4479	1S¹	4.5312	1	4.5312	1.958	2.1
		2×3.4479	1	6.8958	1S¹	4.5312	1	4.5312	2.7344	2.7
		2×3.4479	1	6.8958	1S¹	8.566	1	8.566	3.8203	3.73
C₂H₃	2P¹	2×2.5403/2	1	2.5403	1S¹	3×4.5312/2	1	6.7968	1.8492	2.0
CH₃	2P¹	2.5403/2	1	1.27015	1S¹	3×4.5312/2	1	6.7968	1.070	1.08
C₂H₅	2P¹	2×2.5403/2	1	2.5403	1S¹	5×4.5312	1	22.656	2.284	2.34
C₂H₅	2P¹	2×2.5403/2	1	2.5403	1S¹	5×4.5312/2	1	5.7968	1.849	1.8
O₂	2P¹	3.1658	4	0.81645	2P¹	3.1658	4	0.81645	0.4082	0.43 ± 0.05
H₂	1S¹	6.396 (only for H₂)	1	6.0390	2P¹	6.396	1	6.396	3.198	3.58
F₂	2P¹	2×3.4557	1	6.9114	2P¹	2×3.4557	1	6.9114	3.4557	3.08 ± 0.1
Cl₂	3P¹	2×2.1365	1	4.273	3P¹	2×2.1365	1	4.273	2.1305	2.38 ± 0.1
	3P¹	2.1365	1	2.1365	3P¹	2.1365	1	2.1365	1.068	1.3 ± 0.4
OH	2P¹	3.1658	1	3.1658	1S¹	4.5312	1	4.5312	1.8979	1.83 ± 0.04
OH	2P¹	3.1658	1	3.1658	1S¹	8.566	1	8.566	2.3644	2.29
H₂O	1S¹	2×8.566	1	17.132	2P²	5.982	1	5.982	4.434	5.0
HO₂	1S¹	4.5312	1	4.5312	2P²	2×5.982	1	2×5.982	3.287	3.04

TABLE 7.3 (Continued)

Species											
ClO	$2P^1$	2.1365	1	2.1365	$2P^1$	3.2658	1	3.2658	$2P^1$	1.2915	2.2 ± 0.5
	$2P^5$	6.7989	1	6.7989	$2P^2$	5.982	1	5.982	$2P^5$	3.182	2.91
1	2	3	4	5	6	7	8	9	10	11	
ClO$_2$	$2P^5$	6.7989	1	6.7989	$2P^1$	2×3.2658	1	6.1316		3.331	3.43
ClO$_3$	$2P^5$	6.7989	1	6.7989	$2P^1$	3×3.2658	1	9.7974		4.014	3.96
Br$_2$	$4P^5$	5.7041	1	5.7041	$4P^5$	5.7041	1	5.7041	$4P^5$	2.852	2.6 ± 0.3
J$_2$	$4P^5$	4.8219	1	4.8219	$4P^5$	4.8219	1	4.8219	$4P^5$	2.411	2.4 ± 0.3
BrO$_2$	$4P^5$	5.7041	1	5.7041	$2P^1$	2×3.2658	1	6.8316		3.045	3.22
JBr	$5P^5$	4.8119	1	4.8119	$4P^5$	5.7041	1	5.7041		2.613	2.7 ± 0.2
SO	$3P^1$	1.9265	1	1.9265	$2P^1$	3.2658	1	3.2658		1.212	1.1 ± 0.1
HS	$1S^1$	8.566	1	8.566	$3P^2$	3.3029	1	3.3029		2.384	2.33 ± 0.01
SF	$2P^2$	3.3029	1	3.3029	$2P^5$	13.463	1	13.463		2.652	2.5 ± 0.5
ScH	$4P^2$	2.7405	1	2.7405	$1S^1$	8.566	1	8.566		2.075	2.04
											2.21
NS	$3P^1$	3.0946	1	3.0946	$3P^1$	1.9265	1	1.9265		1.187	1.3 ± 0.3
PO	$3P^3$	3.7713	1	3.7713	$2P^1$	3.2658	1	3.2658		1.750	1.6
PH	$3P^1$	1.7696	1	1.7696	$1S^1$	8.566	1	8.566		1.467	1.6
PH$_2$	$3P^1$	1.7696	1	1.7696	$1S^1$	2×8.566	1	2×8.566		1.604	1.6
PCl$_2$	$3P^3$	3.7713	1	3.7713	$3P^3S^2$	2×13.299	1	2×13.299		3.344	3.26 ± 0.45
AsH$_2$	$4P^1$	1.4551	1	1.4551	$1S^1$	2×4.5312	1	2×4.5312		1.254	1.26
CN	$2P^2$	4.3554	1	4.3554	$2P^2 2S^2$	15.804	1	15.804		3.419	3.21
SiH	$3P^2$	2.0291	1	2.0291	$1S^1$	4.5312	1	4.5312		1.402	1.42
PbCl	$6P^2$	1.7256	1	1.7256	$3P^1$	2.1365	1	2.1365		0.955	1.0 ± 0.02
PbBr	$6P^2$	1.7256	1	1.7256	$4P^1$	1.7113	1	1.7113		0.869	0.9 ± 0.02
BO	$2P^1 2S^2$	5.982			$2P^1$	3.2658				2.101	2.12
BO	$2P^1 2S^2$	5.982			$2P^2$	5.982				2.968	3.1 ± 0.1

TABLE 7.4   Dependence between Covalence and Vander Waals Radii

Period	$\gamma = \left(\dfrac{n*+1}{n*}\right)$	Atom	r_k (Å)	γr_k (Å)	R_B(Å)
I	$\left(\dfrac{1+1}{1}\right)^2 = 4$	H	0.28	1.120	1.10
		B	0.80	1.800	1.75
		C	0.77	1.733	1.70
II	$\left(\dfrac{2+1}{2}\right)^2 = 2,25$	N	0.70	1.575	1.50
		O	0.66	1.485	1.40
		F	0.64	1.440	1.35
		Si	1.11	1.974	1.95
III	$\left(\dfrac{3+1}{3}\right)^2 = 1,778$	P	1.10	1.956	1.90
		S	1.04	1.849	1.85
		Cl	1.0	1.778	1.80
		Ga	1.25	2.017	2.0
		Ge	1.24	2.001	2.0
IV	$\left(\dfrac{3,7+1}{3,7}\right)^2 = 1,6136$	As	1.21	1.952	2.0
		Se	1.17	1.888	2.0
		Br	1.20	1.936	1.95
		Sn	1.40	2.188	2.20
V	$\left(\dfrac{4+1}{4}\right)^2 = 1,5625$	Sb	1.41	2.203	2.20
		Te	1.37	2.141	2.20
		I	1.35	2.109	2.15

In particular, such values of P_E-parameters were successfully implemented for the evaluation of mutual solubility of components in metal systems [15, 16].

In this investigation the following formulas are used as calculated equations to evaluate the affinity for atom electrons (Table 7.2):

$$P_E = \frac{P_0}{kR(n*+1)} \tag{14}$$

and

$$P = \frac{P_E}{N} \approx \chi \tag{15}$$

where N – number of registered valence electrons, k – distance coefficient, usually $k = 1$, but sometimes $k = 2$.

N equals the amount of all valence electrons for the system second period, and for other periods only external valence electrons are considered. In general, P-parameter has a physical sense of the average energy value of one valence electron divided by (n^*+1).

It is demonstrated that for some elements the difference in the reference data corresponds to different valence states of the atom.

Examples: Fe (I, II, III), Co (I, II, III), Cu (I, II), etc. The results of the calculations of χ for binary systems carried out by the principle of adding the reciprocals of component P-parameters are given in Table 7.3.

$$\frac{1}{\chi} \approx \frac{1}{P_c} = \frac{1}{P_1} + \frac{1}{P_2} . \tag{16}$$

In general, such calculation results satisfactorily match the reference sources given in [1, 2]. But Sommer [3, 17] used χ_c values that principally differ from these ones. The analysis of χ_c values in Ref. [17] demonstrates (the table is not available) that these magnitudes can be evaluated through the initial values of P_0-parameters not considering (n^*+1) by the formula:

$$P_E = \frac{P_0}{kR} \approx \chi$$

where P_0 is the parameter of atom external valence electrons.

The correctness of such approach is confirmed by the comparison of P_E-parameters calculated using covalence and Vander Waals radii.

It is known that the electron energy, if there are no other electrons on the orbital, depends only on $(Z^*/n^*)^2$, where Z^* – nucleus effective charge. Equaling the Eqs. (12a) and (12b) and taking to square of n^* and (n^*+1), we have:

$$r_k \left(n^* + 1 \right)^2 - R_d n^{*2} \rightarrow R_d - \left(\frac{n^*+1}{n^*} \right)^2 r_{R} , \tag{17}$$

where r_k – covalence radii, R_B – Vander Waals radii.

The correctness of the Eq. (17) is confirmed by the calculations given in Table 7.4.

7.2 CONCLUSIONS

1. Energy of the affinity for electron (χ) by Refs. [1, 2] equals numerically the average value of P_E-parameter for one valence electron divided by (n^*+1), where n^* – effective main quantum number.

2. Not taking into account the coefficient $(n*+1)$ the values of P_E-parameters of external valence electrons corresponds to χ_c – by the data of Sommer [17].

Covalence and Vander Waals radii are connected with simple dependence through the coefficient $\left(\dfrac{n*+1}{n*}\right)^2$.

KEYWORDS

- **Affinity for electron**
- **Covalence and Vander Waals radii**
- **Lagrangian equation**
- **Spatial-energy parameter**

REFERENCES

1. Energies of Breaking off Chemical Bonds, Potentials and Affinities for Electron, Reference-Book Kondratyev, V. I. (Ed.) (1974). M Nauka, 534p.
2. Properties of Inorganic Compounds, Reference Book, Efimov, A. I., et al. (1983). L Chemistry, 392p.
3. Tables of Physical Quantities, Reference Book edited by Kikoin, I. K. (1976). M Atomizdat, 1008p.
4. Rubin, A. B. (1987). Biophysics Book 1, Theoretical Bio physics, M Higher School, 319p.
5. Aring, G., Walter, J. Z., & Kimbal, J. (1948). Quantum Chemistry, M, I. L., 528p
6. Korablev, G. A. (2005). Spatial-Energy Principles of Complex Structures Formation, Netherlands, Brill Academic Publishers and VSP, 426p (Monograph).
9. Fischer, C. F. (1972). Average-Energy of Configuration Hartree-Fock Results for the Atoms Helium to Radon, Atomic Data, *4*, 301–399.
8. Allen, K. W. (1977). Astrophysical Magnitudes, M. Mir., 446p.
9. Waber, J. T., & Cromer, D. T. (1965). Orbital Radii of Atoms and Ions, J. Chem. Phys., *42(12)*, 4116–4123.
10. Clementi, E., & Raimondi, D. L. (1963). Atomic Screening Constants from, S. C. F. Functions, 1, J. Chem. Phys. *38(11)*, 2686–2689.
11. Clementi, E., & Raimondi, D. L. (1967). Atomic Screening Constants from, S. C. F. Functions, 1, J. Chem. Phys. *47(14)*, 1300–1307.
12. Korablev, G. A., & Zaikov, G. E. (2006). Energy of Chemical Bond and Spatial-Energy Principles of Hybridization of Atom Orbitals, J. Appl. Poly. Sci., *101(3)*, 5, 2101–2107.
13. Clementi, E. (1965). Tables of Atomic Functions, J. B. M. S. Res. Develop, Suppl., *9(2)*, 76.
14. Batsanov, S. S., & Zvyagina, R. A. (1966). Overlap Integrals and Problem of Effective Charges, Novosibirsk, Nauka, 386p.
15. Korablev, G. A. (2010). Exchange Spatial Energy Interactions, Izhevsk, Publishing House, Udmurt University, 530p (Monograph).

16. Korablev, G. A., & Vorobyev Yu P. (1986). Evaluation of Mutual Solubility of the Components of Double Metal Systems, Bulletin of the Academy of Science of the USSR, Metals, *3*, 212–215.
17. Sommer, A. H. (1972). Photo Emission Materials (Translated from English), M Mir.

CHAPTER 8

KINETICS OF MOLECULAR DESORPTION IN A SPHERICAL NANOPORE

M. G. KUCHERENKO, T. M. CHMEREVA, A. D. DMITRIEV, and D. V. STRUGOVA

CONTENTS

8.1   INTRODUCTION

Correct description of the kinetics of processes involving electron-excited molecules is a topical task in view of the development of molecular electronics and photonics. In particular, it is possible to probe nanostructures using the reaction of cross-anni-hilation of electron excitations localized on oxygen molecules and immobile lumi-nophor molecules, which is accompanied by delayed fluorescence [1–3]. As a result of the interaction between an oxygen molecule and immobile luminophor occurring in an excited triplet state, the O_2 molecule passes to an excited singlet state while the luminophor molecule is inactivated. Then, the singlet oxygen molecule can interact with the still unquenched triplet center, which results in its passage into the first ex-cited singlet state that is fluorogenic. The kinetics of this reaction in porous matrices is significantly influenced by the character of mobile reactant motions in a potential field of the pore walls. Namely, the lateral diffusion of oxygen molecules can be interrupted by their desorption's into the gas phase within the pore volume. For this reason, in the course of processing of a time-resolved luminescence signal for extracting information on the system parameters, it is necessary to use mathematical models that take into account peculiarities of the migration of oxygen molecules.

Previously, we have developed [4–6] a theoretical approach according to which the desorption is taken into account in the equations of surface kinetics by intro-ducing the probability $W(t)$ of the occurrence of an excited mobile reactant in the near-surface region of a pore at an arbitrary moment of time t. For determining $W(t)$, the desorption of oxygen was considered in Refs. [4–6] either as the free-radical diffusion in a solid-wall potential or as the passage via a potential barrier according to the Kramers theory [7], if the potential energy of an oxygen molecule in the pore has a barrier-well shape.

In this chapter, the process of oxygen desorption is treated as radial diffusion in a two-well potential that can be formed in the pore in the presence of some filler. For this reason, the kinetics of oxygen desorption is described using an analytical solution of the Fokker-Planck equation with a two-well model potential that is com-posed of parabolic branches and the infinitely high wall. It is established that the intensity and kinetics of the cross-annihilation delayed fluorescence of luminophor molecules are sensitive to changes in the probability of the presence of excited O_2 molecules in the near-surface region of a pore.

8.2   THEORETICAL MODEL

The Fokker-Planck equation for the probability density $g(r, t)$ of finding an excited oxygen molecule at distance r from the pore center at an arbitrary moment of time t for a centro symmetric potential field $V(r)$ has the following form:

$$\frac{\partial g(r,t)}{\partial t} = D \frac{1}{r^2} \frac{\partial}{\partial r} r^2 \left[\frac{\partial}{\partial r} + \frac{1}{k_B T} \frac{\partial V(r)}{\partial r} \right] g(r,t), \tag{1}$$

where D is the coefficient of radial diffusion of oxygen molecules, k_B is the Boltzmann constant, and T is the temperature of the system.

The boundary condition on the pore surface is expressed as zero total diffusion flux:

$$\left[\frac{\partial g(r,t)}{\partial r} + \frac{1}{k_B T} \frac{\partial V(r)}{\partial r} g(r,t) \right]_{r \to R} = 0 , \tag{2}$$

where R is the pore radius.

The initial condition corresponds to a delta-function-shaped distribution of the probability density that reflects the fact of excited oxygen molecule generation at certain point x at the initial moment of time:

$$g(r,0) = \frac{\delta(r - \xi)}{4\pi \xi^2} \tag{3}$$

Since a Boltzmann distribution of the probability density must be attained with the time, another boundary condition is

$$g(r,\infty) = A_0 \exp\left[-\frac{V(r)}{k_B T} \right], \tag{4}$$

where A_0 is a constant factor that is determined from the normalization condition

$$4\pi \int_0^R g(r,t) r^2 dr = 1 ,$$

which must be valid at any moment of time.

A solution of Eq. (1) can be found in the following form:

$$g(r,t) = \frac{u(r)}{r} \exp\left[-\frac{V(r)}{2k_B T} \right] \exp\left[-\lambda^2 D t \right], \tag{5}$$

where l is the eigenvalue of the Fokker-Planck operator. Substituting Eq. (5) into Eq. (1) leads to the following equation for the $u(r)$ function:

$$\frac{\partial^2 u(r)}{\partial r^2} + \left\{ \frac{1}{2k_BT}\frac{\partial^2 V}{\partial r^2} - \left(\frac{1}{2k_BT}\frac{\partial V}{\partial r}\right)^2 + \frac{1}{rk_BT}\frac{\partial V}{\partial r} + \lambda^2 \right\} u(r) = 0 \tag{6}$$

Taking into account Eq. (2) and the fact that probability density $g(r, t)$ at $r = 0$ is limited, we obtain the following boundary conditions for the $u(r)$ function:

$$u(0) = 0, \quad \left[\frac{\partial u}{\partial r} + \frac{1}{2k_BT}\frac{\partial V}{\partial r}u - \frac{u}{r}\right]_{r\to R} = 0 \tag{7}$$

Since the boundary-value problem (6)–(7) with a two-well potential representing a superposition of the Morse and Lennard-Jones atomic potentials has no analytical solution [5], we replace the real potential by the following model:

$$V(r) = \begin{cases} V_0 + k_1 r^2/2, & 0 \le r \le r_0, \\ k_2\left(R^2 - r^2\right)/2, & r_0 < r < R, \\ \infty, & r \ge R, \end{cases} \tag{8}$$

where $k_1 = 2(V_1 - V_0)/r_0^2$, $k_2 = 2V_1/\left(R^2 - r_0^2\right)$, $V_1 - V_0$ is the depth of the central well, V_1 is the depth of the near-wall well, and R is the pore radius. With this model potential, Eq. (6) separates into two equations. In the $[0, r_0]$ segment, we have

$$\frac{\partial^2 u_1}{\partial r^2} + \left\{ \frac{3k_1}{2k_BT} + \lambda^2 - \frac{k_1^2 r^2}{4k_B^2 T^2} \right\} u_1 = 0,$$

While in the $[r_0, R]$ segment we obtain

$$\frac{\partial^2 u_2}{\partial r^2} + \left\{ -\frac{3k_2}{2k_BT} + \lambda^2 - \frac{k_2^2 r^2}{4k_B^2 T^2} \right\} u_2 = 0$$

Solutions of these equations, which obey the boundary conditions (7), can be written as follows:

$$u_1(r) = Ar\exp\left[-\frac{k_1 r^2}{4k_BT}\right] F_1(r), \tag{9}$$

$$u_2(r) = C\exp\left[-\frac{k_2 r^2}{4k_BT}\right]\{rF_2(r) + BF_3(r)\}, \tag{10}$$

where

$$F_1(r) = F\left((3 - q_1)/4;\ 3/2;\ k_1 r^2/(2k_BT)\right),$$

and
$$F_2(r) = F\big((3 - q_2)/4;\ 3/2;\ k_2 r^2/(2k_B T)\big),$$

$$F_3(r) = F\big((1 - q_2)/4;\ 1/2;\ k_2 r^2/(2k_B T)\big)$$

are degenerate hyper geometric functions with $q_1 = 3 + 2k_B T \lambda^2/k_1$ and $q_2 = -3 + 2k_B T \lambda^2/k_2$; coefficient B is given by the formula.

$$B = \frac{R F_2'(R) - k_2 R^2 F_2(R)/(k_B T)}{\big(k_2 R/(k_B T) + R^{-1}\big) F_3(R) - F_3'(R)},$$

and $F_2'(R)$ and $F_3'(R)$ are derivatives of the hyper geometric functions at point $r = R$.

It should be noted that, as $r \to R$, function $V(r)$ tends to infinity and, hence, its derivative tends to infinity as well. Therefore, for obtaining a stable solution, it is necessary to use boundary condition (7) with the value of derivative on the left of point $r = R$, that is, to differentiate the corresponding parabolic function (8).

At point r_0, the solution must obey the following relations:

$$u_1(r_0) = u_2(r_0),$$

$$\left[\frac{\partial u_1}{\partial r} + \frac{1}{2k_B T}\frac{\partial V}{\partial r} u_1 - \frac{u_1}{r}\right]_{r \to r_0} = \left[\frac{\partial u_2}{\partial r} + \frac{1}{2k_B T}\frac{\partial V}{\partial r} u_2 - \frac{u_2}{r}\right]_{r \to r_0}.$$

Substituting Eqs. (9) and (10) into these relations leads to the following equation for the spectrum of eigenvalues λ_n of the Fokker-Planck operator:

$$\frac{F_1'(r_0)}{F_1(r_0)} = \frac{r_0 F_2'(r_0) - k_2 r_0^2 F_2(r_0)/(k_B T) - B\big(k_2 r_0 F_3(r_0)/(k_B T) - F_3'(r_0) + F_3(r_0)/r_0\big)}{r_0 F_2(r_0) + B F_3(r_0)} \quad (11)$$

Thus, a solution of Eq. (6) that corresponds to the n-th eigenvalue has the following form:

$$u_n(r) = A_n R_n(r) =$$

$$= A_n \begin{cases} r \exp\big[-k_1 r^2/(4k_B T)\big] F_1(r), & 0 \le r \le r_0, \\[2mm] \dfrac{r_0 \exp\big[\big((k_2 - k_1) r_0^2 - k_2 r^2\big)/(4k_B T)\big] F_1(r_0)}{\big(r_0 F_2(r_0) + B F_3(r_0)\big)} \big(r F_2(r) + B F_3(r)\big), & r_0 < r < R. \end{cases} \quad (12)$$

Now we can write a solution of the boundary-value problem (1)–(4):

$$g(r,t) = A_0 \exp\left[-\frac{V(r)}{k_B T}\right] + \exp\left[-\frac{V(r)}{2k_B T}\right] \sum_{n=1}^{\infty} A_n G_n(r) \exp\big[-\lambda_n^2 D t\big], \quad (13)$$

where $G_n(r) = R_n(r)/r$.

It should be noted that the first term in the right-hand part of Eq. (13), which represents a Boltzmann distribution, corresponds to the Eigen value $\lambda = 0$. Coefficients A_n are determined from the initial condition (3) and expressed by the formula.

$$A_n = \left(G_n(\xi)\exp\left[V(\xi)/(2k_BT)\right] - 4\pi A_0 \int_0^R \exp\left[-V(r)/(2k_BT)\right]G_n(r)r^2dr \right)\left(4\pi\int_0^R G_n^2(r)r^2dr \right)^{-1}$$

Once the analytical solution of the boundary-value problem (1)–(4) is known, we can determine the probability $W(t)$ of finding an excited oxygen molecule in the near-surface region of a pore using the following integral:

$$W(t) = 4\pi \int_{R-b}^R g(r,t)r^2dr,\tag{14}$$

where b is the thickness of the near-surface layer.

The experimentally measured intensity $I_{DF}(t)$ of delayed fluorescence in the system of identical nanopores with triplet (T) electron-excited centers is determined by the following integral [5, 6]:

$$I_{DF}(t) \sim n_T(t)\int_{4\pi} g_\Delta(\vartheta,t)\theta(\vartheta-\vartheta_0)d\Omega,\tag{15}$$

where $n_T(t)$ is the surface concentration of T centers, $\theta(\vartheta-\vartheta_0)$ is the Heaviside theta-function that determines the angular sector free of T-centers on the sphere with angular parameter ϑ_0, and $g_\Delta(\vartheta,t)$ is the distribution function of singlet-excited oxygen molecules relative to the system of T centers in the pore, which in turn depends on the probability $W(t)$ of the occurrence of an excited mobile reactant in the near-surface region (i.e., the absence of desorption of a singlet oxygen molecule) at the given moment of time t [5, 6].

In concluding this section, it should be noted that, in pores with dimensions of several dozen nanometers, it is possible to restrict the consideration to a one-dimensional Fokker-Planck equation and use linear functions for approximating a real two-well potential.

8.3   RESULTS AND CONCLUSIONS

Calculations using the proposed model have been performed for pores with a diameter of 5 nm. In these pores, the effect of the wall potential field is significant over the entire volume. The potential energy of an oxygen molecule in this cavity can be determined by summing the energies of pair interaction between O_2 molecule and each atom of the solid phase. In the continuous limit of the pair interaction energy expressed by the Lennard-Jones formula, an analytical expression for the effective potential $V(r)$ of a spherical surface has been obtained in Ref. [8]. In this case, the potential energy has the form of a narrow near-wall well with a branch that is steeply ascending on approach to the wall. The energy smoothly increases toward the pore center, so that a molecule can occur at any point of the well. This is not a two-well potential. A two-well potential can be formed if the cavity is filled with macroscopic molecules or with surfactant molecules. Then, a barrier term has to be introduced into the expression for the effective potential [8]. As a result, there appear two potential wells -one at the pore center and another at the wall, as is depicted in Fig. 8.1a. In our model calculations, the depth of the central well was ~25 meV and that of the near-wall well was ~90 meV. The potential barrier maximum occurred at $r = $ 4 nm. Figure 8.1b shows a model potential described by Eq. (8), which retains the main parameters of a real two-well potential.

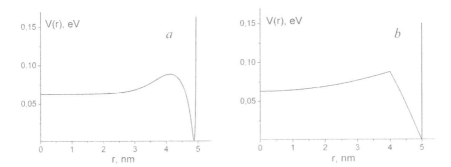

FIGURE 8.1 Effective potential of a spherical solid surface (*a*) in a continuous approximation and (*b*) according to the proposed model.

Figure 8.2 presents the results of calculations of the probability density $f(r, t)$ of finding a singlet oxygen molecule between two spheres with radii r and $r + dr$ at various moments of time:

$$f(r,t) = 4\pi g(r,t) r^2$$

The coordinate dependence of function $f(r, t)$ at the initial moment of time corresponds to a delta-function-shaped distribution with a maximum at 4.5 nm (Fig. 8.2a). This point of singlet oxygen generation was selected so that a distance to the surface would be comparable with the size (~0.5 nm) of an organic luminophor molecule. As can be seen from (Fig. 8.2), function $f(r, t)$ deforms with the time and tends eventually to a form that corresponds to the Boltzmann distribution. The diffusion coefficient of singlet oxygen in these calculations was set to be $D = 10^{-6}$ cm^2/s.

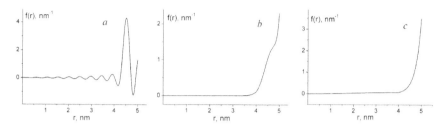

FIGURE 8.2   Profiles of the probability density $f(r, t)$ of finding a singlet oxygen molecule at various moments of time: (a) $t = 0$; (b) $t = 0.5$ ns; (c) $t = 10$ ns.

Figure 8.3 (solid curve) shows the results of calculations of the probability (14) of the absence of desorption of a singlet oxygen molecule. For the comparison, Fig. 8.3 (dashed curve) shows the $W(t)$ function in the case of diffusion of a free oxygen molecule in the potential of the solid wall [5, 6]. The integration in Eq. (14) was performed over a near-wall region of the potential well. As can be seen, the probability of the absence of desorption in the former case tends to a greater constant value (0.85) than that in the latter case (0.5), which is explained by the capture of oxygen molecule in a deep near-wall well and, hence, by a more difficult transition via the potential barrier. In addition, the constant probability level in the case of free diffusion is attained faster than in the case of diffusion in the two-well potential.

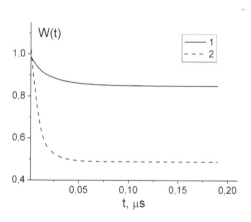

FIGURE 8.3   Temporal variation of the probability of finding a singlet oxygen molecule in the near-wall region of the pore in the case of desorption treated as (curve 1) diffusion in a two-well potential and (curve 2) free diffusion.

Figure 8.4 presents the kinetics of cross-annihilation delayed fluorescence (15) [5, 6], where solid curve corresponds to desorption treated as the diffusion in a two-well potential and the dashed curve corresponds to diffusion in the solid-wall potential.

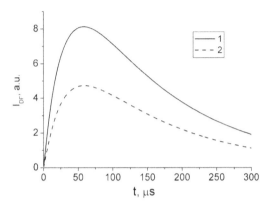

FIGURE 8.4   Kinetics of cross-annihilation delayed fluorescence in the case of singlet oxygen desorption treated as (curve 1) diffusion in a two-well potential and (curve 2) free diffusion.

These calculations were performed for the following parameters: lifetime of the triplet state, $\tau_T = 760$ µs; number of luminophor molecules in the pore, $n_0 = 10$; number of oxygen molecules in the pore, $N_{ox} = 5$; lifetime of singlet oxygen, $\tau_\Delta = 40$ µs; rate constant of the bulk quenching of triplet centers by oxygen in the ground state, $K_\Sigma^{(3)} = 5 \cdot 10^{-15}$ cm³/s. Since the probability of the absence of singlet oxygen desorption in the case of free diffusion is lower than that for diffusion in the two-well potential, the maximum of the delayed fluorescence intensity in the former case is also lower than that in the latter case (Fig. 8.4, dashed curve).

8.4   CONCLUSIONS

Thus, the results of our calculations show that the intensity and kinetics of the signal of cross-annihilation delayed fluorescence are sensitive to changes in the probability of occurrence of excited O_2 molecules in the near-surface region of a pore. For this reason, the desorption of singlet oxygen molecules must be taken into account in processing of the time-resolved luminescence signals that accompany photoreactions involving molecular oxygen in nanoporous materials. These signals contain unique information on the presence of oxygen in pores, the efficiency of oxygen diffusion in pores, the structure of porous media, and features of the interaction of O_2 molecules with the surface of pores.

ACKNOWLEDGMENTS

This work was supported in part by the Ministry of Education and Science of the Russian Federation in the framework of project no. 1.3.06.

KEYWORDS

- **Cross-annihilation delayed fluorescence**
- **Diffusion**
- **Fokker-Planck equation**
- **Nanopores**
- **Surface potential**

REFERENCES

1. Levin, P. P., Costa, S. M. B., Ferreira, L. F. V., Lopes, J. M., & Ribeiro, F. R. (1997). Delayed Fluorescence Induced by Molecular Oxygen Quenching of Zinc Tetra Phenyl Porphyrin Triplets at Gas/Solid Interfaces of Silica and Zeolite, J. Phys. Chem., B, *101*, 1355.
2. Levin, P. P. (2000). Kinetics of Delayed Fluorescence During the Quenching of Eosin Triplet States by Molecular Oxygen on a Porous Surface of Aluminum Oxide, Khim, Fiz, *19*, 100.
3. Kucherenko, M. G. (2001). On the Kinetics of Reaction between singlet Oxygen and Immobile Sensitizers, Khim, Fiz, *20*, 31.
4. Kucherenko, M. G., Chmereva, T. M., & Gun'kov, V. V. (2006). A Model of Excitation Transfer involving Molecular Oxygen on a Solid Sorbent Surface, Khim Fiz, *25*, 95.
5. Kucherenko, M. G., Chmereva, T. M., & Chelovechkov, V. V. (2011). Kinetics of the Cross-Annihilation of Localized Electron Excitations in a Potential Field of Walls of a Porous Nanostructure, Khim Fiz Mezoskop, *13*, 483.
6. Kucherenko, M. G., & Chmereva, T. M. (2012). Pulse Shape of the Cross-Annihilation Delayed Fluorescence of Dyes in Oxygen-Containing Nanoporous Materials, Vestnik Orenburg, Gos., Univ., *9*, 89.
7. Berezhkovskii, A. M., & Zitserman, Yu V. (1995). Elementary Events of Reactions in Solutions, Multi dimensional Kramers Theory, Khim Fiz., *14*, 106.
8. Kucherenko, M. G. (2002). On the Kinetics of Molecular Desorption, Vestnik Orenburg Gos., Univ., *5(15)*, 92.

CHAPTER 9

CALCULATING GAS FLOW IN POLYBUTADIENE PORES DURING PYROLYSIS

A. M. LIPANOV and A. A. BOLKISEV

CONTENTS

9.1 INTRODUCTION

Polybutadiene rubber is one of the most common binder types for condensed solid propellants (CSP); and modeling of processes taking place in its thermal decomposition is an important component of theoretical research on propellants burning laws. Mathematical models of these processes are constructed by Lipanov et al. [1–3]. In this chapter, a numerical method is proposed to solve obtained in these papers systems of differential equations and calculation results are presented.

9.2 PROBLEM STATEMENT

While solid propellant charge is being heated, in condensed phase a thin (about 50 μm) heated layer forms with very steep temperature profile through the charge depth [4]. When surface temperature reaches 650K a process of binder pyrolysis starts [5, 6] which goes with forming of pores in it and porous polymeric skeleton at the surface [7].

Chemical transformations during pyrolysis are described by three irreversible reactions in condensed phase (de polymerization, forming and thermal destruction of polymeric skeleton) according to the model presented in [7], and five reversible reactions in gas phase (reported in [8] to be the main ones with addition of demonization reaction of 1, 3-butadiene into 4-vinylcyclohexene [9]).

Parameters of the flow of emitting gases are determined by means of averaging over finite volumes of the computation domain containing enough number of pores. All the pores are approximated with cylinder of varying cross-section, which axis corresponds to normal to propellant charge surface, with real geometry of pores being accounted for in flow resistance force [1, 2, 10].

To calculate gas velocity and pressure in pores one-dimensional continuity and pressure equations are used

$$\frac{\partial \phi \rho}{\partial t} + \frac{\partial \phi \rho v}{\partial x} = \dot{m}$$

$$\frac{\partial \phi \rho v}{\partial t} + \frac{\partial \phi \rho v^2}{\partial x} + \phi \frac{\partial p}{\partial x} = -R_f$$

where ϕ – porosity, ρ – gas density, v – gas velocity, $\dot{m}$ – mass inflow rate of gases due to binder decomposition, p – pressure, R_f – specific flow resistance force, which is determined by the expression used in Ref. [10]:

$$R_f = A\rho v |v| (b_1 + b_2 / \text{Re}),$$
$$A = 4\phi / d,$$

where A – specific surface of pores, d – pore diameter, Re – Reynolds number, b_1 and b_2 – empirical constants selected so that pressure difference in the pore matches experimental values.

Physical boundary conditions for these equations are [1]:

$$x = x\big|_{T=T_{pyr}} : \quad v = \frac{\rho - \rho_s}{\rho} v_{pyr},$$

$$x = 0 : p = p_\infty, T = T_{sur},$$

where T_{pyr} – temperature at which pyrolysis starts, v_{pyr} – velocity of progression of isotherm $T = T_{pyr}$, ρ_s – binder density, subscript "∞" denotes parameters in the free volume of the engine, subscript "sur" – on the charge surface.

Energy equation is written assuming that kinetic energy of the gases is small comparing to thermal and thermal equilibrium between solid and gas phases establishes instantly [2]:

$$\frac{\partial \overline{\tilde{n}\rho T}}{\partial t} = \frac{\partial}{\partial x} \overline{\lambda} \frac{\partial T}{\partial x} + \dot{Q}_{chem} + \dot{Q}_{heat}$$

with boundary conditions

$$x \to \infty : \quad \frac{\partial T}{\partial x} = 0, \qquad x \to -\infty : \quad \frac{\partial T}{\partial x} = 0,$$

where $\dot{Q}_{chem}$ – heat effect of chemical reactions, $\dot{Q}_{heat}$ – absorbed heat flux, c, ρ and λ – heat capacity, density and thermal conductivity, with bar under the parameters denoting averaging over condensed and gas phases.

Volatile species concentrations are determined form continuity equations, with Fick independent diffusion approximation for gas mixture being used (which is appropriate when diffusion coefficients are not very different or one of the gases is prevalent) while temperature and pressure gradients effects on diffusion are not accounted for:

$$\frac{\partial \phi \rho \omega_i}{\partial t} + \frac{\partial \phi \rho \tilde{v} \omega_i}{\partial x} - \frac{\partial}{\partial x} \phi \rho D_i^M \frac{\partial \omega_i}{\partial x} = \dot{m}_i, \tag{1}$$

where ω_i – mass fraction of i-th species, D_i^M – effective diffusion coefficient of i-th species in the mixture, $\tilde{v}$ – effective flow rate, $\dot{m}_i$ – mass inflow rate of i-th species due to chemical reactions. The boundary condition for species mass fractions on the propellant charge boundary is

$$\frac{\partial \omega_i}{\partial x} = 0.$$

Because of approximate calculation of coefficients D_i^M [11, 12] summing of Eqs. (1) with $\tilde{v} = v$ does not result in continuity equation. To recovery their consistence it is necessary to introduce a correction to the flow rate equal

$$\hat{v} = -\sum_j \omega_i \phi \rho D_j^M \frac{\partial \omega_j}{\partial x},$$

then in Eq. (1) $\tilde{v} = v + \hat{v}$.

The resulting system is supplemented by the equations for porosity

$$\frac{d\phi}{dt} = -\frac{1}{V} \sum_{i \in S} M_i \frac{dn_i}{dt},$$

Power density of heat sources

$$\dot{Q}_{chem} = \frac{1}{V} \sum_i H_i \frac{dn_i}{dt},$$

Absorbed heat flux (Beer-Lambert–Bouguer law)

$$\frac{d\dot{Q}_{heat}}{dx} = -k(x)\dot{Q}_{heat}, \quad \dot{Q}_{heat}(0) = \dot{Q}_0,$$

where V – volume in which chemical reactions take place, M_i – molar mass of i-th species, n_i – amount of i-th species, and summation is over condensed phase species, H_i – enthalpy of i-th species, $k(x)$ – attenuation coefficient, $\dot{Q}_0$ – surface heat flux, and ideal gas state equation.

9.3   NUMERICAL METHOD

All the processes are modeled using the control volumes method on a regular grid. Computation domain parameters are averaged over control volumes and assigned to their centers. To obtain the boundary conditions a system of internal ballistics equations is being solved using zero-dimensional approximation [13, 14].

To account for interactions between the physical processes a splitting approach is used [15]. This allows arranging the calculation in a modular manner, which significantly simplifies programming and makes possible to choose for each of the processes the most efficient method of modeling. With careful calculations sequencing, this method is no less accurate than the globally implicit [16].

The calculation is organized as follows.

1. The time step Dt is estimated from the Courant-Friedrichs-Levi condition for convection.
2. Changes in the species concentrations and pore volume due to the chemical reactions are calculated. Typically, this step requires conducting internal iterations.
3. Viscosity coefficients are determined and the convective transport process is computed using a pressure–Implicit coarse particles method of second order accuracy [14].
4. Diffusion coefficients are determined and the diffusive transport process is calculated using an implicit difference scheme of second order accuracy.
5. Specific heat and thermal conductivity coefficients are determined and the heat transfer process is computed, taking into account the heat effect of chemical reactions, using an implicit difference scheme of second order accuracy.

9.4   COMPUTATION RESULTS

Gas flow in the pore has a wave character (Figs. 9.1 and 9.2), because due to steep temperature profile in the pore (Fig. 9.3) and high activation energy of the first stage of poly butadiene decomposition (Fig. 9.4) [7] gasification of the polymer is uneven in time. Being heated up to the gasification temperature, the upper layers of the polymer quickly decompose, pushing a wave of decomposition products through the pore, but the energy stored in the remaining polymer cannot support this process.

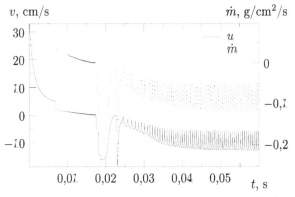

FIGURE 9.1   Flow rate (solid) and mass flow rate (dashed) at the exit of the pore.

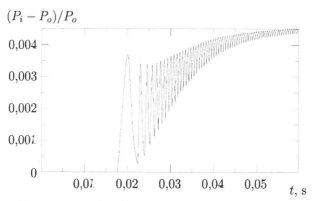

FIGURE 9.2   Relative pressure drop history.

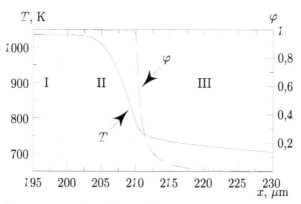

FIGURE 9.3   Temperature and porosity profiles.

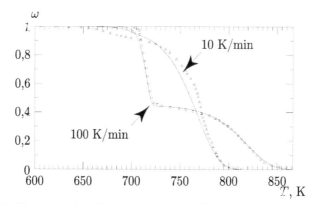

FIGURE 9.4   Thermogravimetric curves of poly butadiene pyrolysis.

Relative pressure drop in the pore increases with its depth and stabilizes in a quasi-stationary mode, when the depth of the pore becomes constant. The gas out-flow rate shows the same behavior. Similar results were obtained in Ref. [17] for a simpler model (except the wave character of the flow, because uniform progression rate of skeleton-polymer boundary was assumed).

Profiles of pressure and flow rate (along with mass flow rate) in a quasi-station-ary mode are presented in Figs 9.5 and 9.6. According to the flow structure, the pore can be divided into a number of segments (starting from the top). At the first of them flow rate increases because of the fact that inflow rate form the walls is not compen-sated by appropriate increase of the pore cross-section. At the second segment the pore cross-section area sharply increases and the flow rate drops. At the end of this segment the flow enters into polymeric skeleton region, passing through the zone of intense decomposition if the initial polymer (it can be seen on the mass flow graph). Flow rate is sharply increases again and then continues to rise because of gas inflow due to polymeric skeleton decomposition until the pore ends; its cross-section at this segment increases non-significantly.

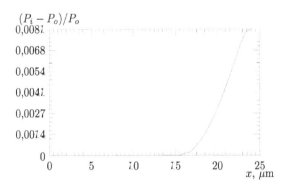

FIGURE 9.5 Profiles of the pressure drop relative to backgrounds.

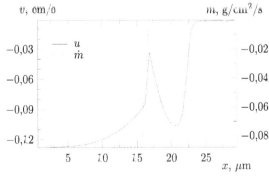

FIGURE 9.6 Flow rate (solid) and mass flow rate (dashed) profiles.

According Fig. 9.7, three stages of the poly butadiene decomposition process may be distinguished. In the first stage (up to about 650K) inert heating takes place. In the second, decomposition of the initial polymer begins with parallel formation of the polymeric skeleton; during that the surface temperature briefly stabilizes at a value about 750K, when the heat flux to the surface is balanced by the endothermic effect of decomposition and heat sink into the polymer depth. Then, the remained near the surface polymeric skeleton heats as an inert substance up to temperature about 1000K, corresponding to the onset of decomposition. Upon reaching the surface temperature about 1100K, heat balance on the polymeric skeleton surface establishes again.

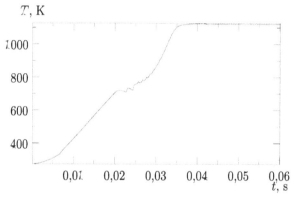

FIGURE 9.7   Surface temperatures history.

Gaseous species residence time in the pore is quite small, so changing of the composition of the out flowing from the pore pyrolysis products due to gas phase reactions is insignificant and corresponding graphs are not shown.

9.5   CONCLUSIONS

Developed is a method to numerically solve the system of non-stationary differential equations describing jointly occurring in the heated region of poly butadiene processes of chemical transformations with pores formations, convection, diffusion and heat transfer. Obtained are profiles of flow rate and pressure in pores, auto-oscillating character of the gases outflow process is revealed.

KEYWORDS

- **Numerical simulation**
- **Polybutadiene**
- **Polymeric skeleton**
- **Pyrolysis**

REFERENCES

1. Kalinin, V. V., Kovalyov, Yu N., & Lipanov, A. M. (1986). Nestacionarnye Process i Metody Proektirovaniya Uzlov RDTT, Moscow, Mashinostroenie, 216p.
2. Bulgakov, V. K., Kodolov, V. I., & Lipanov, A. M. (1990). Modelirovanie Goreniya Polimernyh Materialov, Moscow Himiya, 240p.
3. Lipanov, A. M. (2007). Fiziko-Himicheskaya i Matematicheskaya Modeli Goreniya Smesevyh Tvyordyh Topliv, Izhevsk IPM UrO RAN, 112p.
4. Lipanov, A. M., & Bolkisev, A. A. (2012). On Calculation of a Temperature Field in a Composite Propellant Charge Considering Heterogeneity of its Termo Physical Properties, Chemical Physics and Mesoscopy, *14,* 364–370.
5. Jojic, I., & Brewster, M. Q. (1998). Condensed-Phase Chemical Interaction between Ammonium Per chlorate and Hydroxy-Terminated Poly butadiene, Journal of Propulsion and Power, *14,* 575–576.
6. Caro, R. I., Bellerby, J. M., & Kronfli, E. (2006). Characterization and Thermal Decomposition Studies of a Hydroxy Terminated Polyether (HTPE) Copolymer and Binder for Composite Rocket Propellants, Maintaining Performance and Enhanced Survivability, throughout the Lifecycle, IMEMTS Bristol.
7. On Modeling of Poly butadiene Thermal Decomposition Considering Polymeric Skeleton Formation (2013). Chemical Physics and Mesoscopy, *15,* 236–241.
8. Tsang, W., & Mokrushin, V. (2000). Mechanism and Rate Constants for 1, 3-Butadiene Decomposition, Proceedings of the Combustion Institute, *28(2),* 1717–1723.
9. Duncan, N. E., & Janz, G. J. (1952). The Thermal Dimerization of Butadiene, and the Equilibrium between Butadiene and Vinyl Cyclo Hexene, the Journal of Chemical Physics, *20(10),* 1644–1645.
10. Sulimov, A. A., & Ermolaev, B. S. (1997). Kvazistacionarnoe Konvektivnoe Gorenie v Energeticheskih Materialah s Nizkoj Poristost'yu (Chast' 2), Himicheskaya Fizika, *16(10),* 77–97.
11. Poling, B. E., Prausnitz, J. M. O., & Connel, J. P. (2001). The Properties of Gases and Liquids, New York, McGraw Hill, 803p.
12. Hirschfelder, J. O., Curtiss, Ch F., & Bird, R. B. (1954). Molecular Theory of Gases and Liquids, Wiley, 1249p.
13. Erokhin, B. T. (1991). Teoria Vnutrikamernyh Processov i Proektirovanie RDTT, Moscow, Mashinostroenie, 560p.
14. Lipanov, A. M. et al. (1994). Chislennyj Eksperiment v Teorii RDTT, Ekaterinburg UIF "Nauka," 304p.
16. Oran, E. S., & Boris, J. P. (1987). Numerical Simulation of Reactive flows, Elsevier, 601p.
17. Staggs, J. E. J. (2003). Heat and Mass Transports in Developing Chars, Polymer Degradation and Stability, *82,* 297–307.

CHAPTER 10

POST-CASCADE SHOCK WAVES INFLUENCE ON VACANCY PORES STRUCTURAL TRANSFORMATIONS

A. V. MARKIDONOV, M. D. STAROSTENKOV, and
E. P. PAVLOVSKAYA

CONTENTS

10.1   INTRODUCTION

It is known that influence of the concentrated energy streams on crystal structures by the corpuscular or laser radiation leads to formation of dot defects, thus their concentration can surpass equilibrium concentration considerably. The main mechanism of a relaxation of Meta stable ensembles of such defects is the nucleation and clustering, and, in particular, formation of pores from oversaturated solution of vacancies [1]. Formation of pores can lead to changing of material mechanical properties, and also its volume.

So, for example, it is known that at radiation exposure of a material its swelling that is one of the main reasons for failure of the designs which are operating in the conditions of ionizing radiation is observed. Swelling is caused by development of radiation porosity owing to disintegration of oversaturated solution of vacancies in metal. It is considered that the stationary growth of pores is possible thanks to that dislocation and dislocation loops interact with interstitial atoms because of their bigger mobility in a crystal lattice (the preference phenomenon) more strongly. The driving force of further vacancy pores diffusive evolution is the striving for reduction of a free surface. Thus two tendencies are distinguished: pores coalescence with reduction of their general surface at the invariable volume when psychometric density remains to a constant ("internal" agglomeration), and curing of separate pores with increase in psychometric density ("external" agglomeration) [2]. It is obvious that in the absence of the external squeezing tension, products of nuclear reactions, and also at considerable distance from crystal borders, the collapse of pores is carried out after achievement of some critical size by it when the spherical cavity becomes energetically unprofitable.

Influence of the concentrated streams of energy also can leads to formation of the shock waves, which are called post-cascade [3]. Their emergence is caused by distinction between time of thermalization of nuclear fluctuations in some final area and time of branch of heat from it. As a result of sharp expansion of strongly warmed area almost spherical shock wave is formed. Distribution of shock post-cascade waves can lead to a number of interesting effects: current of a faultless material with hashing of environmental atoms, an abnormal mass transfer, non diffusive processes, phase transformations, sharp increase in number of shifts at atom in volume of a material, even by lack of temperature, necessary for diffusion processes start [3–5], etc.

The purpose of this work is research of post-cascade shock waves influence on processes of pores diffusive evolution.

10.2   EXPERIMENT PART

The phenomena considered in this work are distinguished by the small size of studied areas that complicates direct supervision. Therefore, use of computer modeling methods seems to be most rational. As a method of computer modeling the method of molecular dynamics was chosen because it allows making experiments with set speeds of atoms and comparing dynamics of studied processes to real time. Research was conducted by means of a package of MD-modeling of XMD (Molecular Dynamics for Metals and Ceramics) [6] for which the obvious advantage is the wide set of supported potentials, comparative simplicity of use and openness of initial codes. As potential function of atomic interaction Johnson's potential calculated within a method of shipped atom (EAM) [7] was used. The step of integration equaled 5fs.

Temperature of a calculation cell was set by assignment of casual speeds to atoms according to Maxwell-Boltzmann distribution for the specified temperature. Modeling was carried out at a constant temperature (initial ensemble). For temperature preservation of a calculation cell Andersen's [8] thermostat was used, thus atoms of system experience collisions with certain virtual particles therefore the speed of real particles goes down.

Modeled crystallite of gold had a parallelepiped form. Orientation of crystallite was set as follows: the axis X was directed along the crystallographic direction $<1\bar{1}0>$, an axis Y–length ways $<11\bar{2}>$, and Z – $<111>$. Depending on made experiment along one of the directions free boundary conditions, and along the others periodic was set.

For pores creation in crystal structure the sphere with some radius was set. Then the center of the sphere was combined with one of lattice knots, and all atoms getting to this sphere were removed. After removal of atoms the structural relaxation of a calculation cell before system coming to a state with the minimum energy was carried out.

Waves were created by assignment to the atoms located on border of a calculation cell, speed along H. Close-packed direction was chosen because of existence of energy focusing mechanisms, and the spherical wave is transformed to fragments of the flat waves extending exactly lengthways the close-packed directions [9, 10].

After performance of the set quantity of computer experiment steps the structural relaxation of system followed at 0 K. For visualization of the turned-out structure the visualize of nuclear shifts which represents the lines connecting the initial and final provision of atoms, and also the imposing visualize the close-packed ranks consisting of lines, connecting atoms in three close-packed directions was used.

10.3   RESULTS AND DISCUSSION

For the beginning we will consider process of diffusive evolution of two pores of the equal size. Among all possible options of a relative positioning of pores it is enough to consider of only two: arrangement along the close-packed and not close-packed direction. We will consider the simplest option when pores have the general edge. It is necessary to stipulate that it is a question not of spheres, but of tetra decahedrons as in FCC to a lattice this form of pores has the smallest superficial energy [11]. So, in Fig. 10.1 configurations of pores consisting of 38 vacancies, in a calculation cell are presented.

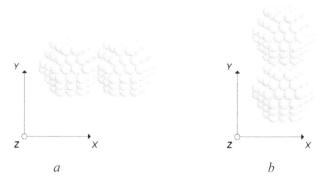

FIGURE 10.1   Double vacancy pores oriented along close-packed (a) and not close-packed (b) directions of crystallite.

At creation of the image in Fig. 10.1, all atoms of the calculation cell which binding energy lies within –4.375 to –3.875eV (the binding energy is equal to an ideal crystal –3.925eV) were removed. Further, at creation of similar images, we will use the same limits of binding energy.

These configurations of vacancies remain stable, up to temperature $0.45 \times T_{mel}$. At higher temperatures reorganization of pores in the uniform complex representing an unfinished tetrahedron of defects of packing is observed. In an ideal tetrahedron of packing defects four sides represent packing defects of subtraction in the planes {111}, and six edges are topmost dislocations with a/6 <110> [12] Burgers' vector. As a rule, in our case violation of symmetry of a tetrahedron consisted in incomplete creation of one of edges or in lack of ideality of one of tops (Fig. 10.2).

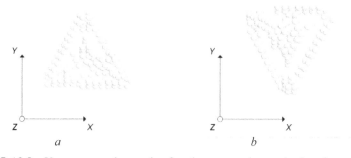

FIGURE 10.2   Vacancy complex made after the structural crystal relaxation, containing pores shown in Fig. 10.1a, at 900 K (a), and pores arranged as shown in Fig. 10.1b, at temperature 600 K (b). The structural relaxation duration was 5000 computer experiment steps.

For studying of influence of the waves extending in a crystal on vacancy pores merge process temperatures at which there is no spontaneous reorganization of pores are of interest. Therefore, we will set the reference temperature of a settlement cell 300 K. For creation of a shock wave to the atoms located on border of a calculating cell, twice big speed was reported to sound speed for gold (s = 3240 m/s [13]).

The conducted research showed the following. Shock waves can cause various vacancy pores transformations, even at a temperature insufficient to start thermo activation of these processes. Thus proceeding structural transformations depend on time intervals through which shock waves in a calculating cell are generated. For task simplification in each separate experiment we will generate waves through equal time intervals. Not the last role is played also by orientation of pores in a calculating cell. So, in Fig. 10.3 configurations of vacancies after generation in the calculating cell containing pores, presented in Fig. 10.1b, ten shock waves through 500 (Fig. 10.3a) and 1000 (Fig. 10.3b) steps of computer experiment are presented.

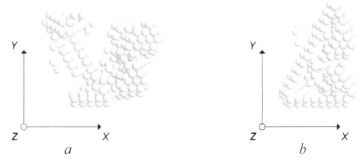

FIGURE 10.3   Vacancy complexes, presented in Fig.10.1b, after ten shock waves through 500 (Fig. 10.3a) and 1000 (Fig. 10.3b) steps of computer experiment.

From Fig. 10.3a, it is visible that the part of vacancies was displaced in the direction of a source of waves, thereby it isn't necessary to speak about a vacancy complex uniform any more, and in Fig. 10.3b the complex of vacancies representing the wrong tetrahedron of defects of packing was created. Thus formed tetrahedron has more correct outlines from a source of waves.

We will consider now the time focused in a calculating cell along the close-packed direction (see Fig. 10.1a). In this case it wasn't succeeded to receive uniform complex as the wave influences a time not evenly. The main restructurings happen in pores located closer to a source of waves, and the wave practically doesn't make influence on the second pore (see Fig. 10.4).

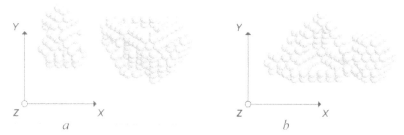

FIGURE 10.4   Vacancy complexes, presented in Fig. 10.1a, after ten shock waves through 500 (Fig. 10.3a) and 1000 (Fig. 10.3b) steps of computer experiment.

According to Fig. 10.4a, shock waves detached part of vacancies from an initial vacancy complex and are displaced to a source of waves.

Thus, in both experiments the detachment and the subsequent shift of part of vacancies from "parental" pores were observed. We will consider this phenomenon in more details. We will generate shock waves in the calculating cell containing a single pore.

The conducted research showed that at the reference temperature of a calculating cell 300 K shock wave initiates reorganization of the single time consisting of 38 vacancies, in a dual tetrahedron of defects of packing. And the part of vacancies is detached from a pore, forming a small tetrahedron, and on a place of a pore the tetrahedron of the bigger size is formed. This detachment of vacancies is initiated by a shock wave therefore the small tetrahedron settles down before a big tetrahedron on a course of distribution of a wave. At 600 K shock wave causes the pore removal, against a course of the distribution. From Fig. 10.5, it is visible that with each subsequent shock wave the pore, which is gradually transformed to a tetrahedron, is displaced in the direction to a source of waves.

FIGURE 10.5   Atoms removal nearby the pore containing of 38 vacancies, depending on shock waves amount (n), gone through calculating cell. The calculating cell temperature is 600 K. Waves are extended from the left to the right according to the figure.

When passing a shock wave through the pore consisting of 337 vacancies, the extreme atoms forming a surface of a pore are beaten out on the distances, exceeding its radius. Therefore, from a pore it is detached about the third part of vacancies, which start being displaced in the direction opposite to the direction of distribution of a wave. Movement is carried out because the wave throws atoms from one pore edge to another. Thus if temperature of a settlement cell makes 300 K, the "parental" time remains on a place (see Fig. 10.6), and in case of temperature equal 600 K it starts collapsing (see Fig. 10.7). Pore destruction, apparently, happens as a result the subsequent waves detach all new vacancies, thereby splitting up the pore. Also it should be noted that at 600 K from a "parental" time the bigger quantity of vacancies after passing of waves is detached, than at a temperature of 300 of K. Some part of the detached vacancies can form fragments of tetrahedrons of packing defects.

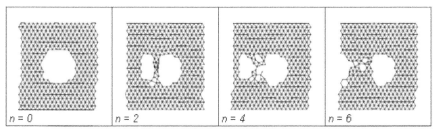

FIGURE 10.6   Structure changes of the pore containing of 337 vacancies depending on the shock waves amount (n), gone through calculating cell. The calculating cell temperature is 300 K. Waves are extended from the left to the right according to the figure.

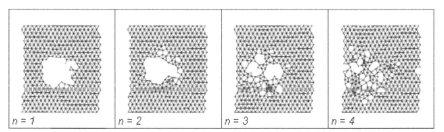

FIGURE 10.7   Structure changes of the pore containing of 337 vacancies depending on the shock waves amount (n), gone through calculating cell. The calculating cell temperature is 600 K. Waves are extended from the left to the right according to the figure.

In summary we will consider experiments with pores of different sizes. Previously it was shown that at orientation of vacancy pores along the close-packed direction of a calculating cell, the wave influence that pore which is closer located to a source of waves. It is obvious that at reduction of the size of this pore impact of a wave on a distant time will increase. At some critical size of a near pore, a wave will cause restructurings of a distant pore, in particular a detachment of vacancies therefore association of pores is possible.

In the Ref. [14], it was shown that the small vacancy congestions representing a dual tetrahedron of packing defects, is the steadiest configuration of vacancies when passing through crystal structure of shock waves, thus waves overcome them with the minimum losses of energy. We will consider such complex in our experiment.

Near the pore consisting of 236 vacancies, we will arrange a complex from 4 vacancies (see Fig. 10.8). Vacancies will be arranged from the frontal party of a time. We will understand that part of a pore, which is turned to an estimated source of waves as the frontal part.

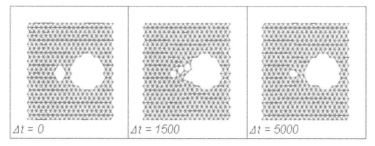

FIGURE 10.8   The calculating cell abstract, containing a pore and a small vacancy accumulation, after the fixed amount of computer experiment made steps Δt. The surface {111} is represented. Waves are extended from the left to the right according to the figure.

In Fig. 10.8, it is shown that as a result of passing of shock waves capture by the pore of small vacancy congestion is really observed. So, at the moment $\Delta t = 1500$ the situation after passing of three shock waves (waves were generated through 500 steps of computer experiment) is presented. The increase in this time interval doesn't lead to desirable results, as it isn't possible to detach vacancy from a pore twice. In the course of a structural relaxation the pore finally absorbs vacancies. By the time of $\Delta t = 5000$ three vacancies from four are absorbed. Thus, it is possible to assume that if sources of waves settle down in a crystal randomly, eventually all small vacancy congestions have to be absorbed by a time.

10.4   CONCLUSIONS

The experiments described in the real work testify that shock post-cascade waves can cause various restructurings of the vacancy pores. So, depending on an arrangement of pores concerning a source of waves, intervals of generation of waves, and also the size of pores, it can be observed either association of pores in a uniform complex, or their crushing on separate components.

Developed provisions can find the application, both in radiation materials science, and when forecasting behavior of the materials operated in extreme conditions. In particular, it is known that the main ways of decrease in radiation swelling of constructional materials consist in change of a structural condition of materials by an alloying, mechanical and thermal processing. It is possible that this work can promote development of a new anti-swelling technique.

ACKNOWLEDGMENTS

Research is executed with financial support of the Russian Federal Property Fund within the scientific project No. 12–02–31135 mol_a

KEYWORDS

- EAM
- Molecular dynamics method
- Stacking fault
- Vacancy pore
- Wave

REFERENCES

1. Mirzoyev, F. H. (2005). Kinetika of a Clusters Nukleation and Formation of Nanostructures in the Condensed Environments, the Collection of Works IPLIT of the Russian Academy of Sciences "Modern Laser and Information and Laser Technologies," 62–78.
2. Cheremskoy, P. G., Slezov, V. V., & Betekhtin, V. I. (1990). Pores in a solid Body, M Energoatomizdat, 376p.
3. Ovchinnikov, V. V. (2008). Radiation and Dynamic Effects, Formation Possibilities of Unique Structural States and Properties of the condensed Environments, Achievements of Physical Sciences, 178(9), 991–1001.
4. Zhukov, V. P., & Boldin, A. A. (1987). Elastic-Wave Generation in the Evolution of Displacement Peaks, Atomic Energy, 68, 884–889.
5. Zhukov, V. P., & Demidov, A. V. (1985). Calculation of the Displacement Peaks in the Continuum Approximation, Atomic Energy, 59, 568–573.
6. XMD–Molecular Dynamics for Metals and Ceramics, [Electronic Resource], Mode of Access, http://xmd.sourceforge.net/about.html.
7. Johnson, R. A. (1988). Analytic Nearest-Neighbor Model for FCC metals, Physical Review, B, 37(8), 3924–3931.
8. Andersen, H. C. (1980). Molecular Dynamics Simulations at constant Pressure and/or Temperature, Journal of Chemical Physics, 72(4), 2384–2394.

9. Chudinov, V. G., Cotteril, R. M. J., & Andreev, V. V. (1990). Kinetics of the Diffuse Processes within a Cascade Region in the Sub-Threshold Stages of, F.C.C. and H.C.P Metals, Physica Status Solidi (a), *122(1)*, 111–120.

10. Garber, R. I., & Fedorenko, A. I. (1964). Focusing of Nuclear Collisions in Crystals, Achievements of Physical Sciences, *83(3)*, 385–432.

11. Palatnik, L. S., Cheremskoy, P. G., & Fuchs, Ya M. (1982). Pores in Films, energoatomizdat M, 216p.

12. Novikov, I. I. (1983). Defects of a Crystal structure of Metals, M Metallurgy, and 232pp.

13. Ultrasounds, Heads, of Golyamin, I. P. (Ed) (1979). "The Soviet Encyclopedia, " M, 400pages.

14. Markidonov, A. V., Starostenkov, M. D. & Obidina, O. V. (2012). Agregatization of Vacancies Initiated by Post-Cascade Shock Waves, Fundamental Problems of Modern Materials Science, *9(4)*, 548–555.

CHAPTER 11

ELECTRONIC STRUCTURE OF PLUMBUM SELENIDE

N. I. PETROVA, A. I. KALUGIN, and V. V. SOBOLEV

CONTENTS

11.1 INTRODUCTION

Binary compounds of the IV–VI group long ago are known as the most perspective materials for the IR photo detectors and laser plants [1, 2]. They differ from the other semiconductors by the many fundamental parameters: very large dielectric permittivity and small energy band, the localization of the long wave transitions at the point L of the Brillouin Zone (BZ) with the p-symmetry of the upper valence band (UVB) and lower conduction band, and other [3, 4]. Therefore, it is not accidentally that their electronic structure and optical properties were theoretically investigated very intensively [3–9]. The interest to this problem after the long interval was greatly increased not only by the actuality of the A^4B^6 crystals for the applications but also by the creating of the modern theoretical methods for the electronic structure calculations [10–14]. They are popular also as the model direct-band binary compounds with the simplest crystal structure of NaCl type with the small energy band $E_g \approx 0.15$–0.3 eV.

The calculation results of Refs. [5–9] and [10–15] are however differs essentially. Moreover they failed the core d-bands, upper conduction bands without the discussion of the spin-orbit importance.

The object of the communication is in the obtaining of the new information about the electronic structure of PbSe and elucidation the nature of the contradictions between the results of the different theoretical calculations.

11.2 CALCULATION METHODS

The electronic structure of PbSe crystal was calculated by the generalized method FPLAPW with the linearized augmented plane waves and exchange-correlation potential in the generalized gradient approximation (GGA) [16–18]. In this paper we used the WIEN2K code [18]. The density of states and the spectra of dielectric function were obtained by the improved method of the tetrahedrons [19]. The maffin-tin spheres were 2.8864 (Pb) and 2.8823 Å (Se), and the lattice parameter of PbSe was equal to $a = 6.1054$ Å.

11.3 RESULTS AND DISCUSSION

The crystal PbSe bands are calculated without the spin-orbit interaction (SOI) and with it in the energy interval $(-20 \div 20)$ eV relatively the maximum of the upper valence band in the L point. For brevity in this communication we discuss the calculation without SOI (Fig. 11.1). The occupied bands may be expansion on the four groups: three UVP with the common width $\Delta E \approx 5.0$ eV (1), the single bands s-type of plum bum (2) and selenium (3) in the intervals of $(-6.3 \div -8.6)$ eV and $(-12.4 \div -13.6)$ eV accordingly, and plum bum bands d-type (4) nearly -17 eV. The width of forbidden band in the L point is equal to 0.37 eV without the SOI and ~0.1 eV with it, the last result is mostly correspondent with the known experimental date for the small temperature [1–4].

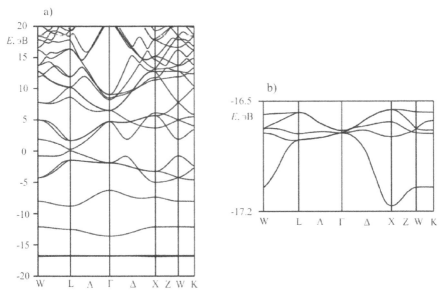

FIGURE 11.1 The PbSe bands calculated without SOI: the valence bands and conduction bands (a) and the core d-bands (b).

The effect SOI is not influenced on the upper valence band V_1 in the points of all directions, except the region nearly G and X points with the SOI splitting of the triple UVB into the V_1 and V_2 the third lower V_3 band (G) at the DE = 0.61 eV, and the splitting the two SOI (V_2, V_3) on the DE = 0.45 eV (X). The degenerated without SOI two lower UVB (V_2, V_3) in the case with SOI are divided nearly with the W, L, G points on the DE = 0.19, 0.22, 0.61 eV, accordingly.

The effect SOI is raised the both bands s-type very small (only on DE H 0.17 eV) and also is divided five bands d-type into the two well separated on the DE H 2.3 eV groups with the three upper and two lower bands in the intervals (−15.52 ÷ −15.87) eV and (−18.21 ÷ −18.39) eV.

The lower conduction band is individual in the points W, L, G and in the wide region nearly two lateral deeper minima in the GX and XWK directions of the BZ. The two upper bands C_2 and C_3 are splitting in the points W and L, and chipped off C_1 in the G point on the DE H 2.1 eV with the splitting of the many other conduction bands.

The energy bands PbSe crystal with SOI which calculated by the different methods for the five points of the BZ in the last 13 year and in the 1975 year are compared in the Tables 11.1 and 11.2. From their analysis it may be the following concludes. The calculated date of [9] in the many years bis to 2000 year was relied as the most correcting for all that it simplification.

The band structure calculations at the last tenth year [10–14] were obtained by the most modern distinguished methods for the five points of BZ. The results mainly are coincide nearly in the correctness of the each calculation for the points L [10, 12, 14], G and X [10–12, 14], accepting the results [15], which different from the other results to the DE = ± 1 eV. The energy bands, calculated by the EPP method [9], are overestimated mainly to the ~0.5–1 eV. The differences between the data of Refs. [10–14] are to the ~1–3 eV. Our results are nearly the data of Refs. [10–12, 14], which calculated only the lower conduction bands to ~5 eV upper the minimum of the conduction bands (LCB) in the contrary to our calculations to ~20 eV with inclusion the many bands. Moreover d-bands were obtained only in the calculations (Table 11.3, Fig. 11.1).

TABLE 11.1   The Energy (eV) of the Conduction Bands of the PbSe Crystal

	Bands	Our data	[9]	[12]	[13]	[14]	[11]	[10]
L	C_1	0.10	0.14	0.11	0.23	0.05	0.28	0.05
	C_2	1.40	1.48	1.51	1.63	1.49	1.65	1.40
	C_3	2.27	2.24	2.35	2.44	2.30	2.27	2.25
	C_4	8.67	10.24	8.44	—	—	—	—
	C_5	10.05	10.95	9.89	—	—	—	—
	C_6	10.35	11.48	—	—	—	—	—
	C_7	10.44	11.53	—	—	—	—	—
	C_8	12.0	12.10	—	—	—	—	—
	C_9	12.1	12.20	—	—	—	—	—
Γ	C_1	3.43	4.10	3.18	3.95	3.18	3.57	3.30
	$C_{2,3}$	5.49	5.90	5.31	5.93	5.31	4.88	5.35
	$C_{4,5}$	6.55	6.24	6.54	6.98	6.54	—	6.55
	C_6	6.65	6.24	6.54	6.98	6.54	—	6.55
	C_7	8.28	7.91	—	—	8.32	—	6.55
	C_8	8.57	8.14	—	—	8.88	—	6.55
	$C_{9,10}$	9.14	9.14	—	—	—	—	6.65

X	C_1	3.77	2.62	3.86	3.84	3.94	—	3.75
	C_2	4.88	6.71	4.53	5.58	4.57	4.47	4.80
	$C_{3,4}$	6.26	8.24	6.03	6.98	5.96	5.29	6.20
	C_5	6.26	9.67	6.03	—	—	5.64	6.20
	C_6	11.45	10.38	—	—	—	—	—
	$C_{7,8}$	11.81	10.52	—	—	—	—	—
K	C_1	3.18	4.05	2.96	3.61	3.5	—	—
	C_2	4.18	4.67	4.08	4.65	3.9	—	—
	C_3	5.49	5.86	5.36	6.05	4.48	—	—
	C_4	6.00	8.10	6.09	6.40	4.64	—	—
	C_5	10.15	10.62	10.00	—	5.97	—	—
	C_6	10.49	10.81	—	—	—	—	—
	C_7	12.41	11.81	—	—	—	—	—
W	C_1	1.80	—	1.73	2.05	1.78	1.92	—
	C_2	4.98	—	4.81	5.34	2.02	4.74	—
	C_3	5.08	—	5.03	5.34	5.02	5.22	—
	C_4	7.57	—	7.32	7.84	—	—	—
	C_5	8.13	—	7.99	—	—	—	—
	C_6	11.95	—	—	—	—	—	—
	C_7	12.76	—	—	—	—	—	—

TABLE 11.2 The Energy of the Valence Bands of the PbSe Crystal in the L, G, X, K, W points with the SOI

	Bands	Our data	[9]	[12]	[13]	[14]	[11]	[10]
L	V_1	0.00	0.00	0.00	0.00	0.00	0.00	0.00
	V_2	1.34	2.05	1.23	1.74	1.25	1.17	1.25
	V_3	1.56	2.19	1.50	1.98	1.45	1.51	1.55
	V_4	8.67	9.05	8.55	9.54	8.71	8.66	8.85
	V_5	12.51	13.38	—	13.26	12.44	12.37	12.55
Γ	$V_{1,2}$	1.62	1.86	1.51	2.10	1.61	1.65	1.60
	V_3	2.26	2.33	2.24	2.67	2.24	2.13	2.20
	V_4	6.21	5.10	6.37	6.51	6.37	6.25	6.35
	V_5	13.57	14.81	—	14.30	—	13.54	13.60

X	V_1	2.96	3.81	2.85	3.72	2.75	2.82	2.90
	V_2	3.43	4.14	3.35	4.10	3.25	3.30	3.35
	V_3	5.00	6.33	4.81	5.70	4.67	4.80	4.95
	V_4	7.27	6.76	7.26	7.91	7.35	7.15	7.35
	V_5	12.10	13.24	—	12.91	12.10	11.90	12.10
K	V_1	2.27	2.76	2.07	2.67	1.73	—	—
	V_2	2.83	3.43	2.63	3.37	2.19	—	—
	V_3	4.60	5.81	4.30	5.35	3.84	—	—
	V_4	7.93	8.00	7.88	8.72	7.54	—	—
	V_5	12.12	13.14	—	12.91	11.74	—	—
W	V_1	0.76	—	0.78	1.20	0.63	0.76	—
	V_2	4.12	—	3.91	4.66	3.92	3.92	—
	V_3	4.33	—	4.19	5.00	4.06	4.06	—
	V_4	7.91	—	7.82	8.41	7.86	7.88	—
	V_5	12.10	—	—	12.50	11.96	11.96	—

Our density of states $N(E)$ for the PbSe crystal with the SOI is on the Fig. 11.2. The upper valence band $N(E)$ with the width $\Delta E \approx 5.0$ eV is caused by p-states of Se and Pb with the contribution of Pb in the several time smaller. This band has very complex structure from the four narrow peaks and two shoulders. Lower are two maxima at the -8.22 and -12.17 eV, caused mainly by the s-states of Pb and Se with the small contribution of the s- and p-states of Se and Pb accordingly. The very narrow peaks $N(E)$ caused with the d-states of Pb are still lower, however the contribution of the d-states Se is smaller nearly to 100 time. The amplitudes of the N(E) peak maxima are bigger nearly 100 time than for the maxima the other states. The spectrum of the conduction $N(E)$ states is consist from the band A (0–5 eV), two narrow peaks (5–7.5 eV), intensive band B (9–13 eV) and many thin intensive peaks in the energy range E > 13 eV. The first band A is caused mainly by the p-states of the both compound components with the visible contribution of the d-state Se and small contribution of the s-state Se in the starting part of band. In consist very thin intensive peak at ~4.9 eV and still the three peaks. The band B with the maximum at ~11.5 eV is consisted from the 5–6 thin peaks. Spectrum of the vacant states $N(E)$ in the energy E > 8 eV mainly caused by the d-states of the both crystal components.

The SOI consideration is caused the splitting of the core $N(E)$ d-band into the two groups with $D_{so}(d) \approx 2.3$ eV, narrowing the triplet structure of the wide occupied band in the interval $(-1.5 \div -5)$ eV and the doublet bands of none occupied states in the intervals $(1 \div 3)$ and $(5 \div 8)$ eV. The complex structure of the occupied states $N(E)$ PbSe were known in the form of the maxima or shoulders of the energy lower than maximum UVB ~1.9, 2.5, 3.2, 3.7, 5.1 eV [20], ~1.5, 2.2, 2.8, 3.3, 4.5 eV [21] with the common width of the valence band DE H 5.0 eV and the splitting energy of the d-bands DE H 2.2 eV, were established by the photoemission spectra. Our calculated data are well in accordance with these experimental results.

The reflectivity spectra of PbSe has very intensive band with the maximum at ~3 eV, several weak peaks in the range 0.5 to 12 eV [4] and the triplet very weak band with R ≈ 0.03 eV in the range 18 to 24 eV [22]. Theoretical curve $e_2(E)$ mainly consists from the one intensive and wide band with the maximum at ~3 eV [3] or from the relatively small band at ~2.5 eV [10]. The calculated spectrum e2(E) PbSe by the energy and intensively is in agreement with the date [15]. This band is caused by the superposition of the many inter band transitions in the BZ volume, especially the transitions between the UVB and LCB. The curve $e_2(E)$ PbSe consists a little visible bands in the range 0.1 to 6 eV due to the similar symmetry of states in the many points of the directions BZ for the UVB and LCB, and also due to absence the parrallelity of the pair energy bands, which must be for the forming the intensive band. Therefore, in the case of PbSe and other compounds of A^4B^6 group it must be carefulness analysis and calculation of the transition field oscillators.

TABLE 11.3 The Energy of the Core d-bands of the PbSe Crystal in the L, G, X, K, W Points with the SOI

Bands	L	Γ	X	K	W
V_6	15.643	15.661	15.618	15.634	15.645
V_7	15.646	15.661	15.658	15.664	15.662
V_8	15.684	15.663	15.769	15.744	15.750
V_9	18.186	18.193	18.185	18.193	18.196
V_{10}	18.194	18.193	18.215	18.213	18.214

Very small reflective intensively PbSe in the range E > 15 eV is caused by the identical d-type symmetry of the core d-bands and conduction bands upper that the UCB bottom by the E > 6 eV.

The reflective intensity of the main semiconductors calculated by means the bands [4, 23] in the range 10 to 23 eV are greater by the 10–100 times, than the experimental data, and also for the PbSe [22]. This peculiarity is evidenced about the non-correctness off the theoretical calculations of the transitions oscillator's fields to the upper conductance bands.

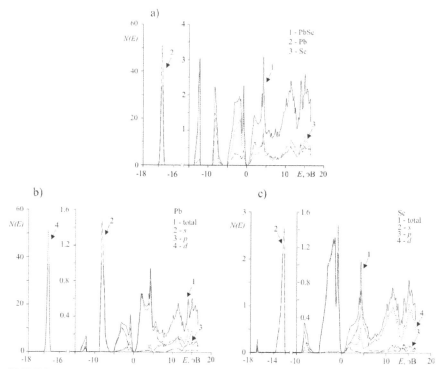

FIGURE 11.2 The spectra $N(E)$ PbSe without SOI: the common (a) and partial contributions of Pb (b) and Se states (c).

KEYWORDS

- **Bands**
- **Brillouin zone**
- **Core levels**
- **Density of states**
- **Energy bands**
- **Plumbum selenide**
- **Spin-orbit interaction**

REFERENCES

1. Ravitch, Yu I., Ephimov, B. A., & Smirnov, I. A. (1968). Methods of the Semiconductor In-vestigations at the Applying to the Plumbum Chalcogenides PbTe, PbSe, PbS, M Science, (In Russian).

2. Baryshev, N. S. (2000). The Properties and applying of the narrows band Semi Conductors, Kazan UNI Press, (In Russian).

3. Dalven, R. (1973). Electronic Structure of PbS, PbSe, PbTe, Sol. St. Phys., *28,* 179–224.

4. Sobolev, V. V. (1981). The Intrinsic Energy Bands of the A^4B^6 group, Kishinev, Science, (In Russian).

5. Lin, P. J., & Kleinman, L. (1996). Energy Bands of PbTe, PbSe, PbS, Phys. Rev. *142(2),* 478–489.

6. Rabii, S. (1968). Investigation of Energy Band Structures and Electronic Properties of PbS, PbSe, Phys. Rev. *117(3),* 801–808.

7. Herman, F., Kortum, R. L., Ortenbutger, J. B., & Van Dyke, I. P. (1968). Relativistic Band Structure of GeTe, SnTe, PbTe, PbSe, PbS, I Phys., (Paris), 29, Col. C4, Suppl., *(11–12),* c62–c77.

8. Kohn, S. E., Yu, P. Y., Petroft, Y., Shen, Y. R., Tsang, Y., & Cohen, M. L. (1973). Electronic Band Structure and Optical Properties of PbTe, PbSe, PbS, Phys. Rev. B, *8(4),* 1477–1488.

9. Martinez, G., Schluter, M., & Cohen, M. C. (1975). Electronic Structure of PbSe, PbTe, Phys. Rev. B, *11(2),* 651–659.

10. Albanesi, E. A., Okoye, C. M. J., Rodriguez, C. O., Blanca, E. L. P., & Petukhov, A. G. (2000). Electronic Structure of PbSe, PbTe, Phys. Rev. B, *61(24),* 16589–16595.

11. Lach-hab, M., Papaconstantopoulos, D. A., & Mehl, M. J. (2002). Electronic Structure of PbS, PbSe, PbTe, I Phys., Chem., Sol., *63(4),* 833–841.

12. Zhang, Y., Ke, X., Chen, Ch., Yang, J., Kent, P. R. C. (2009). Termo dynamic Properties of PbTe, PbSe, PbS, Phys. Rev. B, *80(2),* 024304(12).

13. Svane, A., Cristensen, N. E., Cardona, M., Chantis, A. N., van Schilfgaarde, M., & Kotani, T. (2010). Quasiparticle GW Calculations for PbS, PbSe, PbTe, Phys. Rev. B, *81(24),* 245120 (10).

14. Peng, H., Song, J. H., Kanatzidis, M. G., & Freeman, A. J. (2011). Electronics Structure of Doped PbSe, Phys. Rev. B, *84(12),* 125207(13).

15. Ekuma, Ch E., Singh, D. J., Moreno, J., & Jarell, M. (2012). Optical Properties of PbTe and PbSe, Phys. Rev. B, *85(8),* 085205(7).

16. Perdew, J. P, Burke, S., & Ernzerhof, M. (1996). Generalized Gradient Approximation Made Simple, Phys. Rev. Letters, *77(18),* 3865–3868.

17. Kohn, W., & Sham, L. J. (1965). Self-Consistent Equations Including Exchanges and Correlations Effects, Phys. Rev. *140(4),* A. P., 1133–1138.

18. Blaha, P., Schwarz, K., Madsen, G. K. H., Kvasnicka, D., & Luitz, J. (2001). WIEN2k, an Augmented Plane Wave + Local Orbitals Program for Calculating Crystal Properties, Techn. Univ. Wien, Austria, IS BN, 3–9501031–1, 2.

19. Blöchl, P. E., Jepsen, O., & Andersen, O. K. (1994). Improved Tetrahedron Method for Brillouin-Zone Integrations, Phys. Rev. B, *49(23),* 16223–16233.

20. Grandke, Th, Ley, L., & Cardona, M. (1978). Angle-Resolved UV PE Structures of PbX, Phys. Rev. B, *18(8),* 3847–3871.

21. Hinkel, V., Haak, H., Mariani, C., Sorba, L., Horn, K., & Cristensen, N. E. (1989). Bulk Bands of PbSe, PbTe, Phys. Rev. B, *40(8),* 5549–5556.

22. Martinez, G., Schluter, M., & Cohen, M. L. (1975). Synchrotron Radiation Measurements of PbX, Sol., St., Commun., *17(1),* 5–9.

23. Sobolev, V. V. (2012). The Optical Properties and Electronically Structure of the Non Metals, *I,* Introduction to the Theory, Moscow Izhevsk, IKI.

CHAPTER 12

SIMULATION OF POLLUTANTS GENERATION IN THE COMBUSTION CHAMBER OF THE GAS TURBINE POWER PLANT

E. V. SAFONOV and M. A. KOREPANOV

CONTENTS

12.1   INTRODUCTION

Recently, much importance is given not only efficiency characteristics of power plants using as a source of energy burning process, but also their environmental performance. In this regard, the design of new power plants special attention given to the processes of formation and decomposition of pollutants.

The main components of the polluting combustion power plants using hydrocarbon fuels and air as the oxidant gas are carbon monoxide CO and nitrogen oxides NO_X [1, 2]. At the same time, despite the fact that the kinetics of formation and decomposition of these substances in the flame has been well studied; the problem of modeling these processes remains, in particular due to the complexity of solving systems of equations of chemical kinetics.

12.2   PROBLEM STATEMENT

Gas dynamic path scheme of GTPP, running on natural gas, is presented in Fig. 12.1. Air from the compressor flows around the combustion chambers wall cooling it and entering in a secondary combustion zone and mixing through the perforations. In the two-component injector the air is mixed with natural gas, and enters the combustion zone. After mixing with secondary air the combustion products fed to outlet from the combustion chamber to the turbine. The degree of mixing with secondary air determined by the required temperature level at the turbine no more than 900 °C.

Fuel, as noted above, is natural gas, whose composition was adopted as follows (mass fractions): $CH_4 = 0.9712$, $C_2H_6 = 0.0111$, $C_3H_8 = 0.0027$, $N_2 = 0.0121$, $CO_2 = 0.0029$. Air composition was adopted as follows (mass fractions): $N_2 = 0.7553$, $O_2 = 0.2314$, $Ar = 0.0128$, $CO_2 = 0.0005$.

In the combustion chamber air is supplied with stagnation temperature 726 K after compression and a flow rate of 1.1 kg/s. Fuel supply was carried out at a temperature of 298 K and a total rate of 0.012 kg/s. At the inlet, after the compressor, the pressure was set 0.5 MPa.

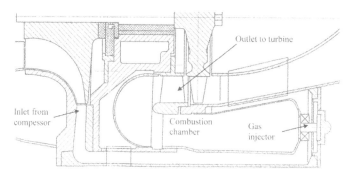

FIGURE 12.1   Gas dynamic path scheme of GTPP.

Due the axisymmetric of gas flow path, and to simplify the calculation the approach with non-stationary ideal mixing reactor [1] was used. Mathematical model of a moving non-stationary ideal mixing reactor and method for its solution are described in detail in Refs. [3, 4]. The main features of this mathematical model are the following assumptions:
- in primary zone near the nozzles formed equilibrium combustion products;
- secondary air is instantaneously evenly distributed over the reactor volume;
- heat transfer of the combustion chamber wall with the products of combustion and cooling air is described using the criteria equations of forced convection [5].

12.3   RESULTS AND DISCUSSION

Numerical simulation was run for two design options: the original version (Fig. 12.1) and with longitudinal fining of the outer wall of the combustion chamber. Finning of outer wall of the combustion chamber leads to an increase in heat exchange with the incoming combustion air, that lead to increase temperature of combustion air on one hand, and decrease temperature of combustion chamber wall on other hand.

Figure 12.2 shows graphs of changes in the concentrations of CO and NO along the length of the combustion chamber. The graphs show that in the combustion zone pollutant concentration are large enough, due to the component ratio close to the stoichiometric and, as a consequence, the high temperature combustion products. In the area of 0.13 m is approached portion of the secondary air, resulting in the concentration of pollutants falling sharply due to dilution air. While carbon monoxide burns and it concentration becomes negligible less than 10^{-6}. It should be noted that the original model has initially low concentration of CO that is associated with a higher initial coefficient of excess oxidant, caused by lower temperature of the primary air with the same geometry of the feed path. On a plot of gas dynamic path from 0.14 m to 0.18–0.20 m nitrogen oxidizes by thermal mechanism described by Zeldovich [6]. Further minor variation in the concentration of NO is due to the mixing of secondary air on residual path of gas flow of the combustion chamber.

However, it should be noted that in both cases the calculated mass fraction of NO tends to its equilibrium value at the current temperature.

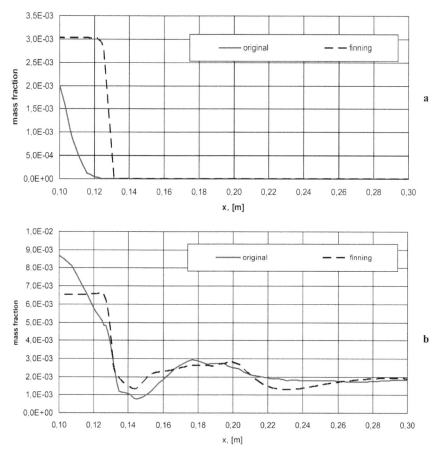

FIGURE 12.2 Changing the mass fractions of CO (a), and NO (b) along the length of the combustion chamber.

12.4 CONCLUSIONS

The studies showed that the main pollutants in combustion products of gas turbine power plants are nitrogen oxides, NO_x, the formation of which is on the thermal mechanism described by Zeldovich [6].

KEYWORDS

- **Burning**
- **Combustion chamber**
- **Gas turbine power plant**
- **Pollutants**

REFERENCES

1. Alemasov, V. E., Dregalin, A. F., Kryukov, V. G., & Naumov, V. I. (1989). Mathematical Modeling of High Temperature Processes in the power Plant, Moscow Nauka, 256p (in Russian).
2. Bowman, C. T. (1975). Kinetics of Pollutant Formation and Destruction in Combustion, Prog., Energy Comb Sci., *1*, 33–45.
3. Korepanov, M. A. (1997). Mathematical Modeling of the Thermal Processing of Solid Propellant Elements dis Candidate, Tehn., Sciences Izhevsk, 162p (in Russian).
4. Korepanov, M. A. (2008). Mathematical Modeling of Chemically Reacting Flows, Chemical Physics and Mesoscopy, *10(3)*, 268–279 (in Russian).
5. Heat and Mass Transfer Theory, Textbook for Universities, Leontev, A. I. (Ed.) (1979). Moscow Higher School, 495p.
6. Zel'dovich, Ya B., Sadovnikov, Ya P., & Frank-Kamenetsky, D. A. (1984). Oxidation of Nitrogen in Combustion, Selected Works, Chemical Physics and Hydrodynamics, Moscow, Nauka, 309–318.

CHAPTER 13

THE EXCITON SPECTRA OF STRONTIUM SULFIDE

V. V. SOBOLEV, VAL V. SOBOLEV, and D. A. MERZLYAKOV

CONTENTS

13.1 INTRODUCTION

Binary compounds IIA–VI group of the alkaline-earth elements IIA (Ca, Sr, Ba) are slightly investigated. They are crystallizing in the cubic lattice NaCl-type and highly perspective as the phosphors with excitation by the γ-ray for radiation dosimetry and as multicolor electroluminescent films. The mono crystals are obtained by a floating melt-zone method.

It follows from the absorption spectra analysis of the SrS crystal in the range 4.1 to 4.3 eV at T= (4÷370) K that the long wavelength edge absorption is caused by the forbidden indirect transitions which in accordance with the theoretical calculation were parent aged from the upper valence band maximum in Γ point into the lower conduction band minimum in the X point [1].

The reflectivity spectra of the SrS cleave were investigated at 77 K in the range 4.5–5.7 eV using the synchrotron radiation.

Our communication is devoted to the complex investigation of the long wavelength edge intrinsic absorption of strontium sulfide with the aim to establish new information about the optical properties and electronic structure of the crystal.

13.2 CALCULATION METHODS

It is generally accepted that the most complete information of the optical properties is contents in the 15 fundamental optical functions [4, 5]: reflectivity (R) and absorption (α) coefficients; index of refractive (n) and absorption (k); imaginary (ε_2) and real (ε_1) parts of the dielectric permittivity ε; real (Re ε^{-1}, Re(1+ ε)$^{-1}$) and imaginary parts (–Im ε^{-1}, –Im(1+ ε)$^{-1}$) of the inverse dielectric functions ε^{-1} and (1+ ε)$^{-1}$; integral function of the combined density of states $J(E)$, which is equal to $\varepsilon_2 E^2$ at the constant intensity of the transitions with the universal factors; effective quantity of the valence electrons $n_{eff}(E)$, participated in transitions to energy E which calculated by the four, methods with the known spectra of ε_2, k, –Im ε^{-1}, –Im(1+ ε)$^{-1}$; effective dielectric permittivity ε_{eff} and other. All these functions are mutually bonded but each function has the self-nature. The physical signification and their interconnection are direct obtained from the general Maxwell equation.

Usually known in the wide energy region only the experimental reflectivity spectrum. Therefore, the spectra of optical function complex are calculated by the special program using Kramers-Kronig interrelations and analytical formula of bonding between functions [3, 4]. The applied calculation methods were completely attracted [3, 4] and discussed [5, 6].

13.3 RESULTS AND DISCUSSION

Our calculations of the spectra of the optical functions complex for the strontium sulfide are obtained using the experimental reflectivity spectra of 2 K (4.5–5.7 eV) and 77 K (4–40 eV) [1, 2].

Experimental reflectivity spectrum in the energy range 4 to 6 eV contains the two doublets (№№1–4, Table 13.1, Fig. 13.1a) which analogs are visible in the spectra of other optical functions (Fig. 13.1b-d, Table 13.1). The maxima in the transverse bands are coincided at R, σ, ε_2, k, α, $\varepsilon_2 E^2$ but for the $\varepsilon_1(E)$ and $n(E)$ they are displaced in the lower energy by ~5–20 meV. Naturally the maxima in the longitudinal band transitions are consisted in the spectra of the volume ($-Im\ \varepsilon^{-1}$) and surface ($-Im(1+\varepsilon)^{-1}$) characteristic electron energy loss with the displacing into the higher energy by the longitudinal-transverse energy splitting E_{lt}=2–7 meV.

The most magnitude were obtained for the band maxima №2 at ~0.47 (R), 17.5 (ε_2), 2.6 (k), 1.3×10^6 cm^{-1} (α), 400 ($\varepsilon_2 E^2$), 10^{16} s^{-1} (σ), 0.18 ($-Im\ \varepsilon^{-1}$), 0.15 ($-Im\ (1+\varepsilon)^{-1}$) and №1 at ~17 ($\varepsilon_1$), 4, 2 ($n$). The absorption coefficient of the most long wavelength band maximum is equal to ~10^6 cm^{-1}, and for the doublet №3, 4 at ~0.4×10^6 cm^{-1} after the exclusion the continuum phone. The large absorption coefficients are evidenced that the all four bands are caused by the dipole allowed direct transitions. It follows from the three $n_{eff}(E)$ spectra that the $\varepsilon_2(E)$, $k(E)$ and $-Im\ \varepsilon^{-1}$ spectra in the energy region to 6 eV are caused by only 0.28, 0.12 and 0.0057 valence electrons SrS from the all 8 in the elemental cell, that is, 1/29, 1/67 and 1/1400 correspondingly. Therefore, the longitudinal components are less be excited relativity the transverse analogs by ~50 times.

The doublet long wavelength band $\varepsilon_2(E)$ is lightly separated by the simplified model of symmetrical contours on the two bands with the areas S_1~12.2 and S_2~17.3 eV, correspondingly which equal to the oscillators strength (the transition probability) whose the equality to the universal multiplier factor (Fig. 13.1b, dotted curves 5, 6 and 7). It may be lightly to estimate the areas of the two-doublet bands №3, 4 (S_3~2.9, S_4~1.1 eV) using the extrapolation of the continuum phone inter band transitions in the energy range 4.9 to 5.5 eV by the curve 8.

It follows from $n_{eff}(\varepsilon_2)$ curve that the magnitude of n_{eff} on the short wavelength tails of the bands №1 and №2 is equal to ~0.029 and 0.075 correspondingly, that is, these band are caused by the 0.029 and 0.046 parts from the all quantity of the valence electrons. These estimate data are equal to oscillator strength of the transitions bands №1 and №2. Their ratio (~1.58) is equal to ratio of the areas on the $\varepsilon_2(E)$ curve (~1.42) with the correction by ~10% which is evidenced of the high correctness of both our estimates for the intensively of the transition bands №№1, 2 SrS crystal.

The maximum at ~4.90 and shoulder at ~4.80 eV of the absorption polycrystalline SrS at 77 K [7] are well evidenced with our calculated data for the α (E) by the energy.

The energy of the spin-orbit splitting of the upper valence band SrS is equal to Δ_{so}~96 meV, which with the high accuracy is coincide with equal energies of the splitting of the both doublets on the optical functions. Therefore, with the large probability it may to propose that doubletons of both bands SrS is caused by the effect of spin-orbit splitting.

The doublets №№1, 2 and №№3, 4 of the SrS reflectivity spectra are caused by the excitions in X and Γ points correspondingly.

The upper valence band and lower conduction band SrS theoretically [8] have the simple at X point and very complex at Γ point structures. More over the energy Δ_{so} at the Γ and X points of many crystals are highly different. The both noted peculiarities in our obtained spectra of the optical functions are not visible.

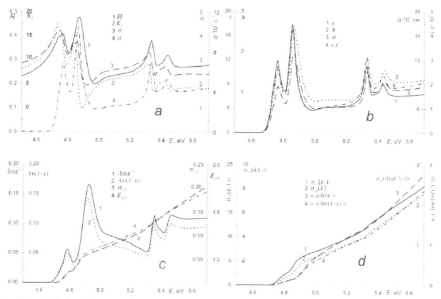

FIGURE 13.1 The experimental reflectivity spectra SrS at 2K [1] (1), calculated spectra ε_1 (2), n (3), σ (4) (a), ε_2 (1), k (2), α (3), $\varepsilon_2 E^2$ (4) (b), $-Im\ \varepsilon^{-1}$ (1), $-Im\ (1+\varepsilon)^{-1}$ (2), $n_{eff}(\varepsilon_2)$ (3), ε_{eff} (4) (c) and $n_{eff}(\varepsilon_2)$ (1), $n_{eff}(k)$ (2), $n_{eff}(-Im\ \varepsilon^{-1})$ (3), $n_{eff}(-Im\ (1+\varepsilon)^{-1})$ (4) (d).

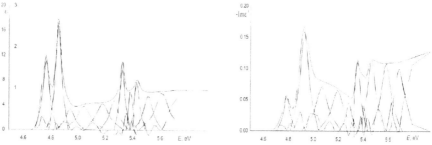

FIGURE 13.2 The decomposition of the ε_2 and $-Im\ \varepsilon^{-1}$ spectra into the elemental components.

TABLE 13.1 The Energy (eV) of the Optical Functions Maxima of the SrS Crystal

N	R	ε_1	n	σ	k	ε_2	α	$\varepsilon_2 E^2$	$-Im\ \varepsilon^{-1}$	$-Im(1+\varepsilon)^{-1}$
1	4.76	4.73	4.73	4.77	4.77	4.77	4.77	4.77	4.79	4.79
2	4.87	4.85	4.84	4.86	4.86	4.87	4.87	4.86	4.93	4.92
3	5.34	5.32	5.31	5.33	5.33	5.34	5.34	5.33	5.36	5.36
4	5.44	5.43	5.42	5.44	5.44	5.44	5.44	5.44	5.47	5.47

TABLE 13.2 The Decomposition Parameters of the ε_2 and $-Im\ \varepsilon^{-1}$ of the SrS Crystal

E		H		Imax		S		$S(\varepsilon_2)/S(-Im\ (1/\varepsilon))\times100$
ε_2	$-Im\ \varepsilon^{-1}$	ε_2	$-Im\ \varepsilon^{-1}$	ε_2	$-Im\ \varepsilon^{-1}$	ε_2	$-Im\ \varepsilon^{-1}$	
4.77	4.79	0.06	0.06	11.00	0.06	0.947	0.005	189.400
4.86	4.93	0.05	0.08	17.00	0.16	1.357	0.020	67.850
4.95	-	0.08	–	3.35	–	0.393	–	–
5.05	5.07	0.11	0.11	3.75	0.07	0.644	0.012	53.667
5.19	5.20	0.11	0.11	4.25	0.07	0.729	0.011	66.273
5.27	5.29	0.08	0.05	4.70	0.05	0.551	0.004	137.750
5.33	5.36	0.04	0.06	10.10	0.11	0.649	0.010	64.900
5.39	5.41	0.04	0.03	6.20	0.04	0.389	0.002	194.500
5.44	5.47	0.05	0.08	7.70	0.12	0.544	0.014	38.857
5.52	5.59	0.08	0.09	5.50	0.11	0.645	0.016	40.313
5.62	5.69	0.12	0.09	6.00	0.10	1.100	0.014	78.571

13.4 CONCLUSIONS

The complex of the optical fundamental function spectra of the strontium sulfide was in the first time calculated at 2 K in the 4.5 to 5.7 eV. It were obtained their main peculiarities and parameters including the spin-orbit splitting energy of both doublet bands of $\varepsilon_2(E)$ which equal to oscillator strength or transitions probability wise the universal multiplier factor. The oscillators strength of both components of the long wavelength doublet bands were obtained using $n_{eff}(\varepsilon_2)$ spectra. All the four maxima of the optical function spectra of the excitons of small radius [9].

KEYWORDS

- **Dielectric permittivity**
- **Effective number of valence electrons**
- **Electron energy losses**
- **Oscillator strength**
- **Reflectivity spectra**
- **Strontium sulfide**

REFERENCES

1. Kaneko, Y., & Koda, T. (1988). New Developments in IIa–VIb (Alkaline-Earth Chalcogenide) Binary Semiconductors, J. Crystal Growth, *86(1),* 72.
2. Kaneko, Y., Mozimoto, K., & Koda, T. (1983). Optical Properties of Alkaline-Earth Chalcogenides II, Vacuum Ultraviolet Reflection Region of eV, J. Phys. Soc. Japan, *52(12),* 4385.
3. Sobolev, V. V., & Nemoshkalenko, V. V. (1988). Methods of the Computational Physics in the Theory of Solid State, Electronic Structure of Semi conductors, Kiev Naukova Dumka, 423p.
4. Sobolev, V. V., Alekseeva, S. A., & Donetskich, V. I. (1976). Calculations of the Semi Conductor Optical Functions by the Kramers-Kronig Inter Relations, Kishinev, Shtiinza, 123p.
5. Sobolev, Val V., & Sobolev, V. V. (2004). Fundamental Optical Spectra and Electronic Structure of ZnO Crystals, Semiconductors and Semi metals, *79,* 201
6. Antonov, E. A., Sobolev, Val V., & Sobolev, V. V. (2010). New Method of the Probing Conduction Bands on the Graphite Example, Proc., VII Intern., Conf., Amorphous and Microcrystalline Semiconductors SPb, Polytech., Univ., 153.
7. Saum, G. A. & Hensly, E. B. (1959). Fundamental Optical Absorption in the IIA–VIB compounds, Phys. Rev. *113(4),* 1019.
8. Hasegawa, A., & Yanase, A. (1995). Electronics Structure of SR Monochalcogenides, J. Phys. C. Sol., St Phys (1980), *13(10).*
9. Sobolev, V. V., & Nemoshkalenko, V. V. (1992). Electronic Structure of Solid State in the Range of Fundamental Edge Absorption, Kiev, V. I., Naukova Dumka, 566p.

CHAPTER 14

SIMULATION OF THE INTERACTION OF FULLERENE AND IRON FE (100)

A. V. VAKHRUSHEV, S. V. SUVOROV, and A. V. SEVERYUKHIN

CONTENTS

14.1   INTRODUCTION

Overview of scientific publications showed that studied in detail the interaction of C60 with molybdenum [1], silicon [2], gold [3] and rhenium [4]. Summary of the studies described in the article [1–4] indicates that the most likely mechanism of the interaction with fullerene metal substrate is the C60 adsorption on the substrate surface until the substrate temperature will not reach the critical value. At temperatures above the critical value of C60 molecules are destroyed and released into the adsorbed layer of carbon dissolved in the bulk substrate molybdenum and rhenium or iridium graphitized in case [5]. All of the above works are experimental.

Theoretical studies of the interaction of C60 with platinum Pt (110)–(1 × 2) and gold Au (110)–p (6 × 5) are given in Refs. [6, 7], respectively. In these papers for research method was used density functional theory. Shown that the interaction of C60 with metal substrates leads to a restructuring of the substrate surface and the formation nanouglubleny. In the case of platinum substrate Pt(110) pentagonal ring C60 contact with the substrate is oriented nearly parallel to the substrate surface. The penetration depth of C6–0 into the gold substrate is a one, two atomic layer of the substrate. It has also been found that between fullerene of C60 and the substrate of gold Au (110) are set tightly focused due C–Au.

The results of the above theoretical studies indicate that the interaction of C60 with a metal substrate is possible not only adsorption but also the penetration of fullerene in the body of the substrate.

The aim of this work is modeling of the interaction with the substrate solid-state cluster containing 9 fullerene C60.

14.2   PROBLEM STATEMENT

The substrate is modeled as a solid-state crystal iron Fe (100). Crystal lattice of Fe is body-centered cubic, the lattice constant is 2.87 Å [8]. Estimated cell had periodic boundary conditions.

The initial state of the C60 cluster and Fe substrate is shown in Fig. 14.1. Numerals in Fig. 14.1a denote the number of C60 in the cluster. Projection of the center of mass of the central plane of C60 on x0y the initial time coincides with the origin. Parameter "a" is the distance between the fullerene x0y plane at the initial time, the parameter "h" is the distance between the fullerene and the substrate plane Fe x0z at the initial time.

Each cluster of fullerenes is given equal speed 0.005; 0.01; 0.02 Å/fs, the velocity vector is directed towards the substrate. Distance between the top of the substrate of Fe and cluster of C60 is 11.5 Å. After contact C60 with the substrate, the speed to zero. Scheme for determining the depth of penetration of fullerene substrate is shown in Fig. 14.2.

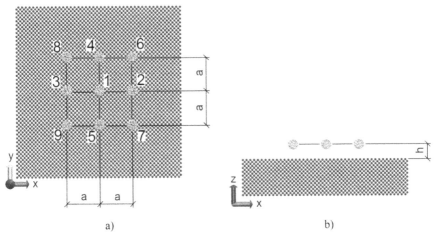

FIGURE 14.1   Cluster of C60 and substrate of Fe.

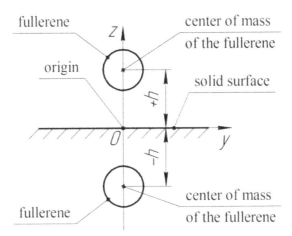

FIGURE 14.2   Determining the distance between the substrate of Fe and the fullerene of C60.

Temperature of the system, substrate of Fe cluster of C60, was 300, 700, 1150 K. Duration modeling of all processes was 30 ps.

To simulate the behavior of a polyatomic molecular system used classical equation of motion of the form [9]:

$$m_i \frac{d^2 \vec{r_i}(t)}{dt^2} = \vec{F_i}, \ 1 \le i \le N,$$ (1)

where i – number of the atom; N – total number of atoms in the system; m_i – mass of the atom; $\vec{r_i}$ – the radius vector atom; $\vec{F_i}$ – the resultant force acting on the atom.

Resultant force is defined as:

$$\vec{F}_i = \frac{\partial U(\vec{r}_1, ..., \vec{r}_N)}{\partial r} + \vec{F}_i^{\,\alpha}, \tag{2}$$

where U – potential energy of the system, $\overline{F_i^{ex}}$ – force determined by interactions with the molecules [9].

The potential energy of the system is a collection of various types of interaction potentials [10]:

$$U(r) = U_b + U_\theta + U_\phi + U_{ej} + U_{VW} + U_{qq} + U_{hb}, \tag{3}$$

where U_b – potential changes in bond length; U_θ – potential changes in the angle between the bonds; U_ϕ – potential changes in the torsion angle connection; U_{ej} – potential changes planar groups; U_{VW} – potential of the van der Waals interaction; U_{qq} – potential electrostatic interaction; U_{hb} – potential hydrogen bonding.

The mathematical form of the specified potentials is described in Refs. [11–15].

When modeling the following potentials was used: changes of length of communication, corner change between communications, changes of the torsion corner. The interaction between the atoms of the crystal lattice of Fe described by Van der Waals potential. Connection between carbon and Fe atoms are not installed.

Modeling the behavior of Fe substrate – cluster of C60 was carried out in the software package LAMMPS. The constants used to calculate the potentials are also given in this software package.

14.3   RESULTS OF MODELING OF C60 CLUSTER AND THE FE SUBSTRATE

Figure 13.3 shows the position of the C60 cluster on the substrate at 30 ps.

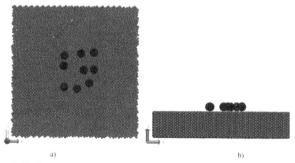

a)   b)

FIGURE 14.3   C60 cluster deposited on Fe substrate.

For research the behavior central cluster fullerene in the vertical plane was investigated behavior of a projection of the center of masses central fullerene on an axis 0z, which is shown in the graph in Figs. 14.4–14.6.

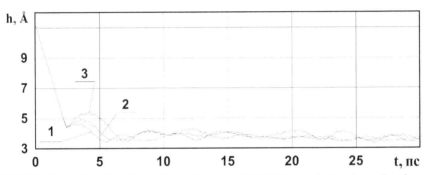

FIGURE 14.4   Changing the position of the central C60 fullerene in the plane x0z when a = 20Å, v = 0.01 Å/fs.

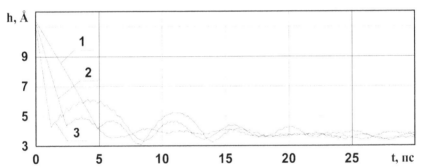

FIGURE 14.5   Changing the position of the central C60 fullerene in the plane x0z when a = 20 Å, T = 700 K.

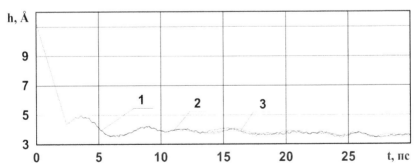

FIGURE 14.6   Changing the position of the central C60 fullerene in the plane x0z when v = 0.01 Å/fs, T = 700 K.

The graph shows that increasing the temperature and velocity of the fullerene is in the first 10–15 ps. It increases the amplitude and period of oscillation in the vertical plane fullerene adsorbed onto the substrate surface. With 25 ps behavior of fullerenes on a substrate is quasi-stationary character. From the graph in Fig. 14.6 show that the distance between the fullerenes does not affect the interaction C60-substrate. Behavior central fullerene cluster horizontal x0y the graphs in Figs. 14.17 & 14.8. To determine the central position of the fullerene Eq. (4) is used.

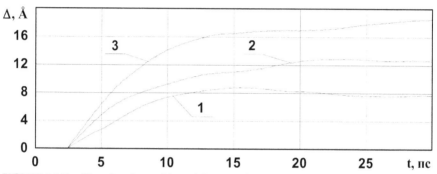

FIGURE 14.7   Changing the position of the central C60 fullerene in the plane x0y when a = 20Å, v = 0, 01 Å/fs.

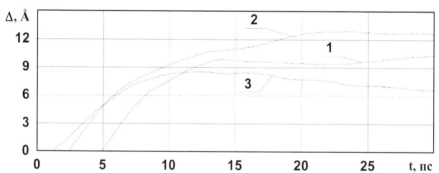

FIGURE 14.8   Changing the position of the central C60 fullerene in the plane x0y when a = 20Å, T=700 K.

From the moment of contact of each cluster fullerenes with the substrate begins to increase the distance between the current and the initial position of the fullerene. Influence of velocity on the behavior of fullerenes in plane x0y is multidirectional. As in the case with the behavior of the fullerene in the vertical plane, the plane x0y behavior does not depend on the distance between the C60 fullerene.

To investigate the behavior of peripheral fullerene cluster relative to the central plane x0y, measured the distance between them:

$$\begin{cases} \Delta_{1-k}(m) = \sqrt{\sum_{n=1}^{2}\left(x_1^n(t_m) - x_k^n(t_m)\right)^2}, \\ x_k^n(0) = X_k^n(0) \end{cases} \qquad (4)$$

where $n = 1, 2$ – an independent coordinate; $m = 1 \dots M$ – the number of time steps, $k = 2, 3, \dots 8$ – number of fullerene cluster; Δ_{1-k} – a distance between the centers of mass of the central and peripheral C60 fullerene; $X_k^n(0)$ – coordinates of the center of mass of the C60 fullerene.

As follows from Fig. 14.9, the behavior of fullerenes relative to the central peripheral is multidirectional, but with 25 ps comes quasi-stationary behavior of the fullerene cluster.

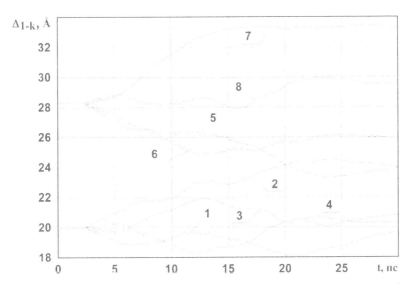

FIGURE 14.9 Changing the distance between the centers of mass of the central and peripheral C60 fullerene in plane x0y when v = 0.01 Å/fs, a = 20 Å and T = 700 K.

At research of cluster deposition on the substrate in the software package LAMMPS, it was established that at a speed equal to the C60 v = 0.02 Å/fs and a temperature of system T = 1150 K, a number of fullerene separation from the substrate, as shown in Table. 14.1 and in Figs. 14.10 and 14.11.

TABLE 14.1   Quantity Not Adsorbing the C60

The distance between the fullerene cluster, Å	15	20	25
Quantity not adsorbing C60 fullerene, numbers.	4	5	3

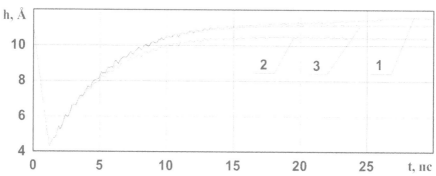

FIGURE 14.10   Changing the distance between the centers of mass of the central and peripheral C60 fullerene in plane x0z when v = 0, 02 Å / fs, T = 1150 K.

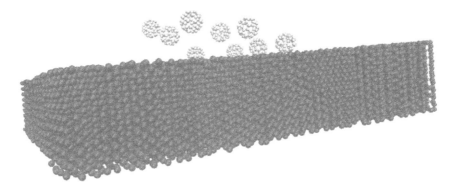

FIGURE 14.11   C60 fullerene not settled on a Fe substrate at a = 20 Å.

The simulation of the C60 cluster interaction with the Fe substrate revealed the most significant parameters affecting the state of the system.

14.4   CONCLUSIONS

The calculation results in the following conclusions:

1.   At any combination of the model parameters, it was observed, at contact of fullerene with a substrate there is fullerene movement on a substrate until they didn't hold steady position. The time required for fullerene that would be in a stable position is about 15 ps.

2. Under certain combinations of speeds of fullerenes and C60 fullerene system temperature is not adsorbed on the Fe substrate surface most likely reason for this is that the temperature fluctuations of the substrate.
3. With 25 ps in the system cluster C60 fullerenes – Fe substrate comes quasi-steady state.
4. Modeling the behavior of the fullerene cluster on the Fe substrate showed that the initial distance between the fullerene does not affect the processes of interaction in the system. This is explained by a sufficiently large distance between the fullerene (15 or more Angstroms). Study the interaction of denser clusters is the aim of further research.

ACKNOWLEDGMENTS

This work was supported by RFBR grant 11-03-00571-a and Programs Basic Research of RAS №12 "Multi-level study of the properties and behavior of advanced materials for advanced friction units" (project 12-T-1-1009).

KEYWORDS

- **Cluster**
- **Depth penetration**
- **Fullerene**
- **Molecular dynamics**
- **Solid**
- **Temperature**

REFERENCES

1. Hang, Xu, Chen, D. M., & Creager, W. N. (1993). Double Domain Solid C60 on Si, (111) 7×7, Phys. Rev. Lett., *70*, 1850–1853.
2. Kuk, Y., Kim, D. K., Suh, Y. D., Park, K. H., Noh, H. P., Oh, S. J., & Kim, S. K. (1993). Stressed C60 Layers on Au, (001), Phys. Rev. Lett., *70*, 1948–1951.
3. Gal, N. R., Rutkov, E. V., Tontegode, Ya A., & Usufov, M. M. (1999). Interaction with the Surface of the C60 (100), Mo, Technical Physics, *69(11)*, 117–122.
4. Gal, N. R., Rutkov, E. V., Tontegode, Ya A., & Usufov, M. M. (1997). Application of Molecular C60 Deep Carbonization of Rhenium in an Ultrahigh Vacuum, Letters to Journal of Technical Physics, *23(23)*, 26–30.
5. Popova, O. I., Glebov, V. A., Glebov, A. V., & Kostyuk, Yu G. (2008). Using Fullerenes to Increase Mechanical Strength of the Composite Magnets, Geteromagnitnaya Microelectronics, *5*, 26–34.

6. Torrelles, X., Langlais, V., Santis, M., Tolentino, H. C. N., & Gauthier, Y. (2010). Nano Structuring Surfaces, Deconstruction of the Pt, (110), (2×1) Surface by C60, Phys. Rev. B, *81*, 041404.1–041404.4.

7. Hinterstein, M., Torrelles, X., Felici, R., Rius, J., Huang, M., Fabris, S., Fues, H., & Pedio, M. (2008). Looking Underneath Fullerenes on Au (110), Formation of Dimples in the substrate, Phys. Rev. B, *77*, 153412.1–153412.4.

8. Kittel, C. (1978). Introduction to Solid State Physics, Moscow, Nauka, 789p.

9. Shaitan, K. V., & Tereshkina, K. B. (2004). Molecular Dynamics of Proteins and Peptides, Oikos, M., 103.

10. Alikin, V. N., Vakhrouchev, A. V., Golubchikov, V. B., Lipanov, A. M., Serebrennikov, Yu S. (2010). Development and Aerosol Nano Technology research, Mashinostroyeniye, 196p.

11. Cantor, Ch, & Schimmel, P. (1984). Biophysical Chemistry in 3 Volumes, Lane from English New York, Wiley, *1*, 336.

12. Andersen, H. C. (1980). Molecular Dynamics Simulations at Constant Pressure and or Temperature, J. Chem. Phys. *72*, 2384–2393.

13. Frenkel, D., & Smit, B. (2002). Understanding Molecular Simulation from Algorithms to Applications, San Diego, Academic Press, 638p.

14. Haile, M. J. (1992). Molecular Dynamics Simulation, Elementary Methods, Inter Science, 386p.

15. Wiley, N. Y., & Nose, S. A. (1984). Molecular Dynamics Methods for Simulation in the Canonical Ensemble, Mol. Phys., *52*, 255–278.

CHAPTER 15

FORMATION, OPTICAL AND ELECTRICAL PROPERTIES OF Ca_3Si_4 FILMS AND $Si/Ca_3Si_4/Si$ (111) DOUBLE HETEROSTRUCTURES

N. G. GALKIN, D. A. BEZBABNYI, K. N. GALKIN, S. A. DOTSENKO, I. M. CHERNEV, and A. V. VAKHRUSHEV

CONTENTS

15.1   INTRODUCTION

"In situ" temperature dependent Hall measurements [1] have shown that a thin Ca_2Si layer grown on 2D Mg_2Si template on Si(111) substrate at (120–130)°C is characterized by 1.02 eV of energy band gap, which precisely corresponds to Ca_2Si quasi particle band structure calculations [2]. Complex study of thin Ca_2Si layers growth at 120 °C and their embedding in silicon has been carried out on Si(111) and Mg_2Si template with different thicknesses [3]. The thin continuous Ca_2Si layer is broken into silicide nanocrystals and embedded in mono crystalline Si substrate in depth up to 20 nm during the Si cap growth at 100 °C. Additional investigations at high substrate temperature showed that at 500 °C a growth of Ca silicide phase with increased Si content [4] occurs directly on the silicon surface. This silicide phase has a composition closed to Ca_3Si_4 and possesses an indirect band gap of 0.63 eV [4] that close to a value obtained from the data of *ab initio* calculations [5]. More detailed investigations of inter band transitions by method of photo reflectance spectroscopy shown [6] that in the film with composition closed to Ca_3Si_4 the strong direct inter band transitions with 0.891 and 0.914 eV are also observed. So, these films can be interested for creation of the photo detectors with high absorption coefficient in the energy range until 1.0 eV. However, thin and thick films with Ca_3Si_4 composition and double hetero structures (DHS) with silicon on their base cap layer were not earlier grown and studied. Since $Si/Ca_3Si_4/Si$ hetero structures are interesting for silicon planar technology for creation of optoelectronic devices, so methods of DHS formation with good crystalline quality, peculiarities of the electronic structure, optical and electrical properties, must be systematically studied.

In this article the morphology, electronic structure, optical and electrical properties of the Ca_3Si_4 films with different thicknesses and $Si/Ca_3Si_4/Si(111)$ double hetero structures grown on their base were studied depending on the thickness of deposited Ca and on method of silicon cap layer growth in DHS.

15.2   EXPERIMENTAL RESEARCH

Thin and thick films of Ca_3Si_4 were grown in the ultra-high vacuum "VARIAN" chamber with a base pressure 2×10^{-10} Torr by reactive deposition epitaxy (RDE) of Ca on Si(111)7×7 substrates. For the creation of double hetero structures Si cap layers were grown atop of Ca silicide films. The deposition rates of Ca ($u_{Ca} = 0.4$– 2.0 nm/min) and Si ($\upsilon_{Si} = 4.7$–4.9 nm/min) were calibrated by quartz microbalance method for all series of experiments. Rectangular Si strips (5×18 mm²) of p-type conductivity with resistivity 45 W cm were used as substrates and Si sublimation sources. The purity of the Si substrates after high temperature annealing at 1250 °C and the concentration of elements in the grown films were determined by Auger electron spectroscopy (AES) and the phase composition of grown films was determined by electron energy loss spectroscopy (EELS) after finishing of growth procedures. The control of phase composition and element concentration in grown films was carried out by AES and EELS after finishing of Ca (Si) deposition and cooling down to room temperature.

Three series of samples were grown during growth experiments. In the first series the films of Ca silicide were grown by reactive deposition epitaxy (RDE) of different thicknesses (3 nm, 30 nm, and 76 nm) of Ca on an atomically clean Si surface (Si(111)7×7) at 500 °C. The 76 nm Ca was deposited at u_{Ca} = 2.0 nm/min and also at 0.4 nm/min rates. In the second and third series by three samples in a series with the same Ca thicknesses and covered by Si cap layer with 100 nm thickness. The samples in series were only differed by a method of the silicon growth: molecular beam epitaxy (MBE) or solid phase epitaxy (SPE) method at temperature of 500 °C.

The morphology of the Ca silicide films and hetero structures were studied by atomic force microscopy (AFM) using Solver P47 in tapping or contact mode. Optical reflectance and transmittance spectra were studied at room temperature on Hitachi U-3010 spectrophotometer and Fourier spectrometer Bruker Vertex 80v in the photon energy range of 0.05–6.2 eV (with integrated sphere at 1.5–6.2 eV). Raman spectra were registered at room temperature on "NTEGRA SPECTRA" with exiting laser wavelength of 488 nm. Low temperature measurements of sample resistance were carried out by two-probe method in the continuous-flow electrical-optical cryostat (RTI Cryo magnetic Systems) with a closed cycle cooling system and automatic registration of constant current and voltage. Thermoelectric measurements were carried out in the temperature range of 20–300 °C in the special vacuum chamber with two heaters (one is for creation of the temperature gradient, and second is for the sample heating up to given temperature), two tungsten probes and two thermocouples apprised to the sample during the measurement procedure.

15.2.1 FORMATION OF CA₃SI₄ FILM ON SI (111)7×7 AND IT ELECTRONIC STRUCTURE

The purpose of this part of the work was a formation of island, thin and thick films of Ca silicide with Ca_3Si_4 composition, which showed semiconducting properties be data of the first investigations [4]. Islands of Ca silicide were observed after deposition of 3 nm Ca on the silicon that was proved by AFM method after the reloading of the sample and also *in situ* due to appearance of the Ca peak with energy of 299 eV on AES spectrum (Fig. 15.1a) and changes of the fine structure of AES Si peak, which characterize the formation of Si-Ca bonds. Formed Ca silicide islands occupy only small part of the surface (smaller 5% by AFM data). The last is also followed from analysis of EELS spectra (Fig. 15.1b). The main contribution in the spectrum gives the bulk silicon plasmon 17.2 eV, but a surface plasmon (10.2 eV) is decreased on amplitude and shifted in the high energy side that confirms a disordering of the silicon surface due to formation of calcium silicide islands. The bulk plasmon is shifted in the small energy side (16.0 eV) and widened from the low energy side that confirm the formation of Ca silicide bulk plasmon with position closed to 14.6–14.8 eV, and formation of the interband transition 6.2 eV, which are usually observed for thick Ca silicide film with the composition of Ca_3Si [4]. So, the formed islands of Ca silicide have a composition closed to Ca_3Si_4. An increase of Ca thickness until to 30 nm resulted in formation of continuous Ca silicide film. This instance is confirmed by EELS spectrum (Fig. 15.1b), on which two peaks dominate: a surface plasmon with 11.2 eV energy and bulk plasmon with 14.8 eV, but also an interband transition at 6.2 eV with increased amplitude characterized to Ca_3Si_4 films [4]. Contributions from bulk and surface plasmons are absent that also confirm the formation of the continuous Ca_3Si_4 film. An increase of deposited Ca thickness up to 76 nm however did not result to formation at 500 °C the continuous Ca_3Si_4 film as was yearly observed in Ref. [4]. Such a behavior character of all system we connect with larger Ca deposition rate (2 nm/min) that preliminary in 5 times larger as compared with Ref. [4] (0.42 nm/min). This results to stoichiometry change in the growing Ca silicide layer across to low silicon diffusion rate from the substrate to the growing Ca silicide film. This conclusion is confirmed by AES data (Fig. 15.1a). The visible intensity of Si peak does not see on the film surface, and Ca peak is shifted in the side of low energies on 1.0–1.5 eV that testifies about non-finished reaction of Ca atoms with silicon. The contribution of the interband transition at 6.2 eV characterized for Ca_3Si_4 [4] is absent on the EELS spectra (Fig. 15.1b). Positions of surface (9.8 eV) and bulk (12.6 eV) plasmons do not also correspond to Ca silicide with composition Ca_3Si_4 [4]. But peaks have visible amplitude and small width that correspond to smooth surface and single-phase of grown silicide. Therefore, the grown film has in the subsurface region a composition reached by Ca. So, for the thickness increase at conservation of a composition and substrate temperature of 500 °C the deposition rate of Ca atoms must be decreased to 0.4–0.5 nm/min.

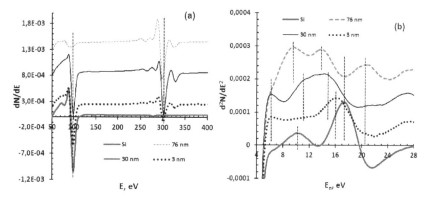

FIGURE 15.1 AES (a) and EELS (b) spectra of atomically clean substrate Si(111) −7*7 and Ca silicide films (Ca_3Si_4) grown on Si substrate by method of reactive epitaxy with different Ca thicknesses.

15.2.2 Si/Ca₃Si₄/Si(111) DOUBLE HETERO STRUCTURE FORMATION

AES and EELS spectra for samples with Ca silicide films covered by silicon layer by MBE method and also spectra for atomically clean silicon surface are presented on Fig. 15.2. It is seen from AES data (Fig. 15.2a) that for all grown samples Ca is observed in subsurface layer with near equal concentration. The fine structure of Si AES peak corresponds to electronic structure of atomically clean silicon that testify about small contribution of silicon peak of Ca silicide in the registered signal. Small regions in the form of pinholes with depth of few nm and lateral sizes of 100–200 nm are observed by AFM data that permit speak about exit of Ca silicide nanocrystallites in the subsurface region during Si growth. Simultaneously on the EELS spectra (Fig. 15.2b) a shift of the surface plasmon in position 8.5 eV is only observed, but the bulk plasmon does not really shifted and widthened. Therefore, one can speak about the creation of double heterostructure, in which nanocrystallites with Ca_3Si_4 composition (from data of Fig. 15.1a) are embedded in the silicon matrix. An increase of Ca thickness up to 30 nm and it growing by silicon resulted to increase of pinhole density on the silicon surface and their lateral sizes up to 200–300 nm.

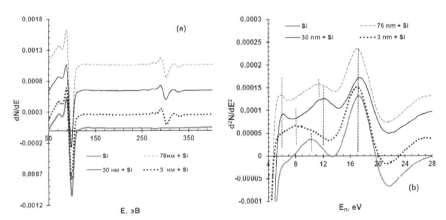

FIGURE 15.2 AES (a) and EELS (b) spectra of atomically clean substrate Si(111) −7x7 and DHS Si/Ca₃Si₄/Si(111) grown on Si substrate by method of reactive epitaxy Ca with different thickness and MBE of silicon on Si(111) substrate.

This is accompanied by appearance of 6.2 eV peak of the interband transition and shift of the surface plasmon in the position of 12.0 eV, but also by broadening of the Si bulk plasmon from the high energy side that testify in favor of Ca_3Si_4 bulk plasmon contribution with energy of 14.6–14.8 eV [4]. An increase of the thickness of deposited Ca up to 76 nm and silicon deposition by MBE method resulted in increase of pinhole area on the sample surface, their lateral sizes (300–500 nm) and depth up to 20–30 nm. An interband transition of 6.2 eV characteristics to Ca_3Si_4 [4] and widened surface plasmon (11.3 eV) and widened bulk plasmon (16.8 eV) indicated about a contribution of Ca_3Si_4 bulk plasmon at 14/6–14.8 eV were appeared on the EELS spectrum. Therefore, a deposition of Si on the Ca silicide film reached by Ca resulted to transformation of this silicide into Ca_3Si_4 under the Si layer.

The third sample series was covered by silicon by method of solid-state epitaxy (SPE) at the same Ca thicknesses. A growth of silicon atop Ca_3Si_4 islands resulted to formation of the continuous silicon film without traces of Ca by AES data in the subsurface region (Fig. 15.3a). At the same time the formation of the interband transition (6.2 eV) characteristic of Ca_3Si_4 [4], small shift of surface plasmon (11.0 eV) and non-changed position of bulk plasmon are observed in EELS spectrum (Fig. 15.2b). So, in this case the double heterostructures (DHS) with embedded Ca_3Si_4 nanocrystallites were also formed. An increase of the Ca thickness up to 30 nm resulted to an appearance of AES signal from Ca (Fig. 15.3a). The pinholes with non-covered surface of silicide were observed by ASM data. The same situation was observed for silicide film with Ca thickness of 76 nm. For both DHS the small shifts of surface and bulk plasmons and small amplitude of 6.2 eV interband transitions were observed in EELS spectra (Fig. 15.3b). It intensity was higher and half-width was smaller for Ca thickness of 30 nm than for silicide film with Ca thickness of 76 nm. But since AES and EELS spectra were registered only after silicon deposition at room temperature (Fig. 15.3a, b), but after the annealing was only registered plan AES spectrum with large peak amplitude for Ca and Si, then the sample annealing resulted to pinhole formation in the silicon cap film with larger area. Therefore, one can speak about worsening of the crystalline quality of grown DHS with an increase of embedded layer of Ca silicide. The silicon films are polycrystalline and consist of grains with 100–200 nm sizes by AFM data.

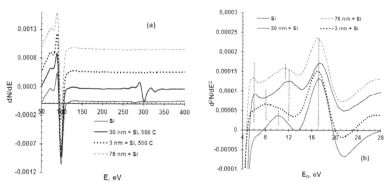

FIGURE 15.3   AES (a) and EELS (b) spectra of atomically clean substrate Si(111) −7x7 and DHS Si/Ca_3Si_4/Si(111) grown on Si substrate by method of reactive epitaxy Ca with different thickness and SPE of silicon on Si(111) substrate.

An analysis of the morphology of grown DHS shown that for the formation of continuous silicon layer at the same substrate temperature the Si thickness must be increased independently from the growth method: MBE or SPE.

15.2.3 OPTICAL END ELECTRICAL PROPERTIES OF CA₃SI₄ AND DHS ON THEIR BASE

Let us consider optical properties of grown Ca silicide films and double hetero structures on their base. Two peaks: main on amplitude at 520 cm^{-1} and the second order peak at 302 cm^{-1} were observed in the Raman spectrum of Si substrate. In Raman spectrum the considerable changes are observed after the growth of Ca$_3$Si$_4$ film deposited with small rate (0. 42 nm/min) as in the Ref. [4]. The group of well-separated peaks at 346, 388, 416 и 453 cm^{-1} with small half-width besides the silicon peak were appeared in the Raman spectrum that testify about good crystalline quality of the film and an absence of the contribution of Ca silicide with another composition. Taking into account the yearly finished analysis of the electronic structure of Ca silicide film grown at the same conditions [4] one can rate the grown Ca silicide film to one type silicide–Ca$_3$Si$_4$, possessing by the reach structure of active phonons in Raman spectrum account for large volume of the unit cell with the hexagonal structure (a=0.8541 nm, c = 1.4906 nm [5]), containing six formula units.

The reach peak structure (0.87, 1.5 and 2.5 eV) is also observed for given DHS (76 nm of Ca) (Fig. 15.4b) that sufficiently well coincide with reflectance spectrum data for DHS sample grown at 500 °C [6]. A sharp increase of the absorbance in the transmittance spectrum (Fig. 15.4c, d) with simultaneous reflectance increase is the main feature of the sample as in a Ref. [6]. Such a behavior is characteristic for plasma reflectance resonance on free carriers well known for semiconductors [7]. So, the Ca silicide film embedded in silicon matrix possesses by high conductivity that coincide with data of Ref. [6]. Peaks of phonon oscillations from the top silicon layer at 609 cm^{-1} and from silicon oxide at 1100 cm^{-1} are seen on the transmittance spectrum in the region of small wave numbers (Fig. 15.4c) that additionally prove the practically full covering by silicon of Ca silicide layer and formation of DHS.

Temperature resistance dependences for clean silicon substrate and Ca silicide films with different Ca thickness on silicon substrate are presented on Fig. 15.5a. Ca silicide films were grown on high resistivity substrate (45 W·cm) for the decrease of shunting effect by silicon substrate. The Ca silicide film of smaller thickness (30 nm) was protected from the oxidation by Si film of amorphous silicon (8 nm) deposited at room temperature. It is seen that the sample resistance on 4–5 orders smaller than silicon substrate resistance in the all temperature region used. The resistance of Ca silicide film doesn't covered by amorphous silicon shows for this an independence from the temperature in range used that testify about high carrier concentration in the grown film and nonmetallic conductivity character in it and it is confirmed by data of optical investigations (Fig. 15.4c, d). The carrier freezing, which occurs in the top Si layer, is observed in the sample with thin Ca silicide film and protecting Si layer at low temperatures (lower than 40 K). At higher temperatures (50–280 K) a contribution of Ca silicide film in the conductivity is crucial as compared with silicon amorphous layer. Temperature measurements of Seebeck coefficient on the Ca silicide film with Ca thickness of 76 nm and on the DHS $Si/Ca_3Si_4/Si$ shown (Fig. 15.5b) that Seebeck coefficient is positive and small (20–50 mkV/K) for both systems in the temperature range of 50–200 °C, and therefore, the majority carriers responsible for generation of thermal electromotive force are holes. The small value of Seebeck coefficient correlates with high concentration of majority carriers (holes) by conductivity data (Fig. 15.5a) and optical transmittance and reflectance spectra (Fig. 15.4c, d). An additional decrease of Seebeck coefficient in the double hetero structure as compared with non-covered Ca silicide film connects with decrease of carrier concentration in an amorphous silicon layer.

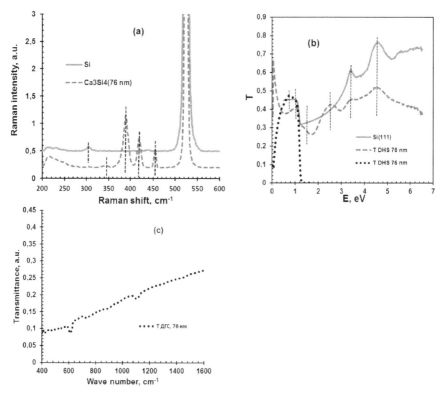

FIGURE 15.4   Raman spectra from mono crystalline silicon substrate and Ca silicide film with Ca thickness of 76 nm (a), optical transmittance (T) and reflectance (R) spectra of DHS with Ca thickness of 76 nm (b) and transmittance spectrum of the same sample (DHS) (c) in the region of small wavelengths (photon energies).

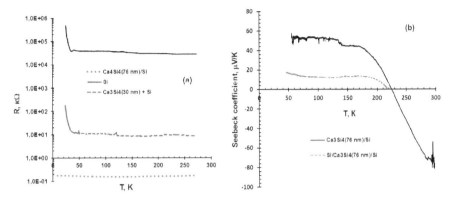

FIGURE 15.5 Temperature dependences of the resistance of silicon substrate (KDB-45), Ca_3Si_4 with Ca thicknesses of 76 nm and 30 nm (a) and temperature dependences of the resistance of silicon substrate (KDB-45), Ca_3Si_4 films with Ca thicknesses of 76 nm and 30 nm (a) and temperature dependences of Seebeck coefficient for Ca_3Si_4 films and DHS Ca_3Si_4 $Si/Ca_3Si_4/Si$ with Ca thickness of 76 nm (b).

15.3 CONCLUSIONS

The formation, optical and electrical properties of Ca_3Si_4 films and double hetero structures with embedded Ca silicide films of different thicknesses and coating continuities have been studied. It was established that in the double hetero structure $Si/Ca_3Si_4/Si$ the nanocrystallites or continuous Ca_3Si_4 layer are formed independently from a thickness of deposited Ca layer. A complex analysis of Raman spectroscopy and optical spectroscopy data has shown that Ca_3Si_4 films are characterized by wide set of Raman active peaks at 346, 388, 416 and 453 cm^{-1} with low width, which proves single-phase of the system and it good crystalline quality. It was shown that this Ca silicide possesses by a high conductivity realized by holes and low values of Seebeck coefficient (20–60 mkV/K). It was established that an increase of the Ca deposition rate at substrate temperature of 500 °C results at Ca thickness increase to growth of Ca silicide enriched by metal, but additional silicon deposition at the same temperature ensures the formation of double hetero structure with maximal thickness of deposited Ca (76 nm).

ACKNOWLEDGMENTS

The work was financially supported by grants of RFBR (No. 13-02-00046_a) and Ministry of Education and Science RF (No 8751).

KEYWORDS

- **Calcium**
- **Calcium silicides**
- **Double hetero structures**
- **Properties**
- **Semiconductor films**
- **Silicon**

REFERENCES

1. Dotsenko, S. A., Fomin, D. V., Galkin, K. N., Goroshko, D. L., & Galkin, N. G. (2011). Physics Proceedia, *11,* 95.
2. Lebegue, S., Arnaud, B., & Alouani, A. (2005). Phys Rev. B, *72,* 085103.
3. Dózsa, L., Molnár, G., Zolnai, Z., Dobos, L., Pecz, B., Galkin, N. G., Dotsenko, S. A., Bezbabny, D. A., & Fomin, D. V. (2013). Journal Matters Science, *48,* 2872
4. Dotsenko, S. A., Galkin, K. N., Bezbabny, D. A., Goroshko, D. L., & Galkin, N. G. (2012). Physics Proceedia, *23,* 41.
5. Migas, D. B., Shaposhnikov, V. L., Filonov, A. B., Dorzkin, N. N., & Borisenko, V. E. (2007). J. Phys. Condens Matter, *19,* 346207.
6. Bezbabny, D. A., Galkin, K. N., Dotsenko, S. A., Galkin, N. G., Zielony, E., Kudrawiec, R., & Misiewicz, J. (2013). Proceedings of APAC Silicides Conferences (July 27–29–2013) Tsukuba, Japan, 28, 13
7. Pancove, J. I. (1971). Optical Processes in Semi conductors, Dover, New York, 386p.

CHAPTER 16

STRUCTURAL AND CHEMICAL TRANSFORMATIONS OF THE MECHANO-ACTIVATED CALCIUM LACTATE IN VORTEX MILL

O. M. KANUNNIKOVA, S. S. MIKHAILOVA, V. V. MUHGALIN, V. V. AKSENOVA, F. Z. GILMUTDINOV, and V. I. LADYANOV

CONTENTS

16.1 INTRODUCTION

Calcium lactate, as well as other carboxylates, belongs to the first-generation products containing no other microelements or vitamins [1]. These products excel due to their well-studied effect on the human body and low cost that makes them affordable for most people. Their disadvantage is low efficiency. Over recent decades, the drug efficiency has been improved through mechanical activation [2]. The goal of this study is to analyze the structural and chemical transformations of calcium lactate observed during mechanical activation in a vortex mill.

16.2 EXPERIMENTAL RESEARCH

The study subject is calcium lactate penta hydrate used to treat diseases related to calcium deficiency. The calcium-lactate mechanical activation was performed in a vortex mill VME-150 (1 and 2 cycles).

The structural and chemical state of the calcium lactate before and after the mechanic treatment was surveyed through X-ray diffractions, XPS and IR spectroscopy.

The physical and chemical properties were analyzed: true density and solubility of calcium-lactate powders, pH and refraction index of aqueous solutions, temperature dependence of aqueous-solution density and capillary viscosity. The chemical and structural state of the aqueous solutions was surveyed through IR spectroscopy.

The true density of the powders was found through the psychometric method (a pycnometer of 5 cm^3). The density of the aqueous solutions was found with an Ostwald pycnometer (of 0.9 cm^3).

The IR spectra of the powders were obtained with an IR Fourier spectrometer FSM 1202 (provided by Monitoring Company (OOO Monitoring)). The powders in question were pressed in tablets with KBr. The XRD analysis of the powders was performed with an X-ray diffraction meter, Bruker D8 Advance, with the use of CuK$_\alpha$ radiation. A solid-state Si (Li) spectrometer, Sol-XE (Bruker), was used as a detector.

The X-ray photoelectron spectra were excited with Mg K$_\alpha$ radiation by an electronic spectrometer ES-2401. The charge was recorded by correlating the hydrocarbons line and the bond energy of 285.0 eV.

The pH was found with a pH-meter, MULTITEST IPL-301, with an indicating glass electrode ESL-43-07. The reference electrode was a silver-chloride electrode. The refraction values were measured with a refract meter IRF-454B2 M. The measurements were performed with transmitted light $(\lambda = 584$ nm) at a temperature of 25 °C. The aqueous-solutions viscosity was measured with a capillary-viscosity meter VPZh-2 (a capillary diameter of 0.37 mm).

16.3   RESULTS AND DISCUSSION

Figure 16.1 shows X-ray diffract grams of the original and mechano-activated calcium-lactate samples. Besides the calcium-lactate lines, the diffractograms have calcium-carbonate lines. Calcium lactate is synthesized through the neutralization of lactic acid with calcium carbonate:

$$CH_3CH(OH)COOH + CaCO_3 \rightarrow [CH_3CH(OH)COO]_2Ca + CO_2\uparrow + H_2O. \qquad (1)$$

The medical product usually contains impurities of lactic acid and calcium carbonate. According to our estimates, the amount of calcium carbonate is 3.25%wt. The diffractograms lines of lactic acid and lactate anions cannot be separated.

After one mechanical-activation cycle, the X-ray lines slightly broaden and lose intensity, as the grain size reduces and the micro strain level grows. Meanwhile, the crystalline structure of the calcium lactate is constant. After two mechanical-activation cycles, the crystalline structure changes (as evidenced by the emergence of additional reflexes) and amorphization processes begin. The calcium-lactate amorphization is also caused by the heating of the original sample up to 115°C (Fig. 16.2).

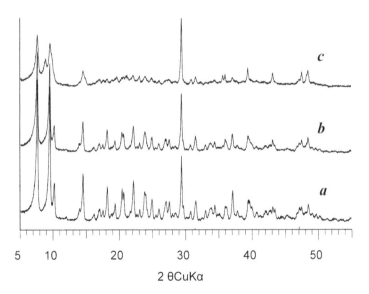

FIGURE 16.1   X-ray diffractograms of the original (a) and mechano-activated calcium lactate (b–1 mechanical-activation cycle, c–2 mechanical-activation cycles).

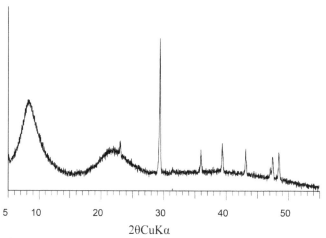

FIGURE 16.2   X-ray diffractograms of calcium lactate after heating up to 115°C.

The IR spectrum of the original calcium-lactate sample is shown in Fig. 16.3. After 1 cycle of calcium-lactate mechanical treatment, the difference IR spectra (Fig. 16.4) demonstrate lower intensity within 3000–3500 cm^{-1} as a result of the dehydration of the original calcium-lactate crystallo hydrate. It is known [3] that there are several hydrated forms of calcium lactate: monohydrate, tri hydrate, pent hydrate. The pentahydrate dehydration begins at 55 °C, runs fast and causes loss of crystallinity. The growing density of the mechano-activated powder (Table 16.1) is caused by calcium-lactate dehydration.

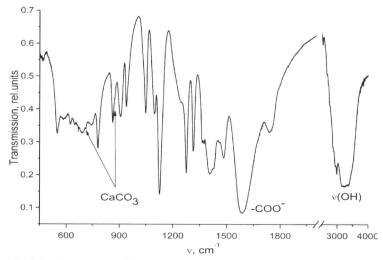

FIGURE 16.3   IR spectrum of the original calcium-lactate sample.

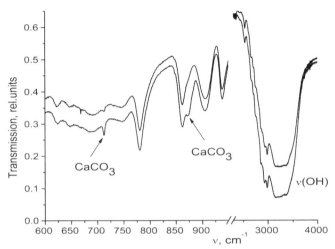

FIGURE 16.4   IR-spectra of powders source (1) and milling (2) calcium lactate.

TABLE 16.1   Physical and Chemical Properties of Calcium Lactate Before and After Mechanical Activation

Sample	Properties		
	True density, g/cm³	Solubility, g/100 g of H₂O	pH of aqueous solution
Original	1.5044	8.36	5.00
1 MA cycle	1.5276	8.40	5.26
2 MA cycles	1.5600	8.40	5.53

The increasing carbonate calcium peaks intensity (at 713 and 875 cm⁻¹) evidences carbonyl-type calcium-lactate decomposition that requires a temperature about 200 °C. This process may be represented as follows:

$$[CH_3CH(OH)COO]_2Ca \rightarrow CaCO_3 + CO_2\uparrow + H_2O. \tag{2}$$

The XPS spectra are given in Fig. 16.5. The most intense line (285.0 eV) refers to the C-H bonds from the CH₃- end group of the lactate anion and from the carbon of the adsorbed hydrocarbons layer. The components with bond energies of 286.2 and 287.4 eV in C1 s-spectrum of the original sample refer to the carbon atoms bonded with the hydroxyl groups in the calcium-lactate molecules of various isomerism types. The first mechanic-activation cycle changes the correlation between the peak intensities of 286.2 and 287.4 eV that may suppose changing correlation of the quantities of isomers. After two mechanical-activation cycles, calcium lactate is available in one structural (isomeric) state. The high-energy peak $E_b = 289.0$ eV refers to the carbon atoms in the carboxyl group of the lactate anion and carbonate anion.

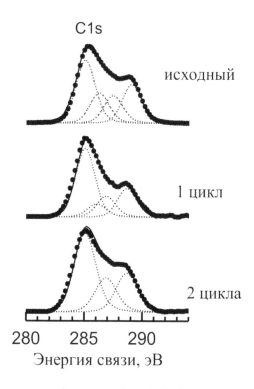

C1s

исходный

1 цикл

2 цикла

280   285   290

Энергия связи, эВ

/исходный – original
1 цикл – 1 cycle
2 цикла – 2 cycles
Энергия связи. эВ – Bond energy, eV/

FIGURE 16.5   C1 s-spectra of lactate on different phase of mechanical activation.

Comparison of IR-spectra of calcium lactate powder and differential spectra of the powders after 1 and 2 cycles of milling shown in the (Fig. 16.6). It is observed the decrease in intensity in the field of stretch vibrations of CH_3 and CH_2 groups (2700–3000 cm^{-1}) due to symmetric and asymmetric vibrations of CH_3 –group (2861 and 2985, respectively) and an increase of intensity of CH_2-groups (peaks at 2855 and 2923 cm^{-1}) vibrations, which is a consequence of the destruction of calcium lactate.

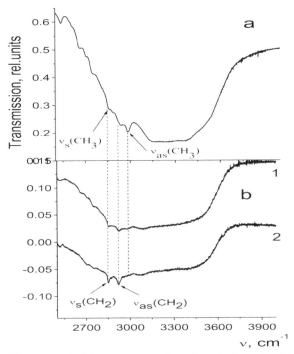

FIGURE 16.6 IR-spectra of calcium lactate powders of: original powders (a) and differential spectra of milling powders (b): 1 – after 1 cycle; 2 – after 2 cycles.

The mechanical activation includes chemical reactions and destruction of the main compound component, calcium lactate; nevertheless, the density, dynamic viscosity and refraction index of the aqueous solutions of the mechano activated medical product powders are just slightly different from those of the original powder solutions. The reason seems that the above chemical reactions involve a minor share of the powders components [4]

The equation $\varphi = (\eta_{rel} - 1)/2.5$ is used to calculate the volume ratio of water molecules within the hydration shells (Fig. 16.7). We see that the ratio of hydration shells reduces as the mechanical activation goes on. The difference for the solutions of the original calcium lactate after 1 mechanical-activation cycle is insignificant, while the hydrate shells of the sample calcium-lactate solution after 2 mechanical-activation cycles demonstrates far less water ratio. It supposes different structural states of the lactate anion before mechanical activation and after it.

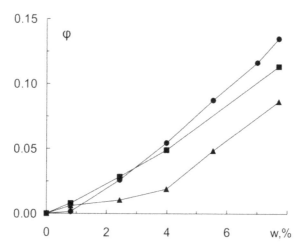

FIGURE 16.7 Concentration dependence of the water volume ratio within the hydrate shells at 26°C: sample product (●), 1 mechanical-activation cycle (■), 2 mechanical-activation cycles (▲).

Table 16.2 gives the volume expansion thermal coefficients of the solutions at 26–40°C and 40–60°C. Obviously, the structural states of the calcium-lactate solutions after 2 mechanical-activation cycles, original lactate solutions and lactate solutions after 1 mechanical-activation cycle are considerably different. The coherence of the calcium-lactate solutions after 2 mechanical-activation cycles is higher than that of pure water.

TABLE 16.2 Volume Expansion Thermal Coefficient of Water and Aqueous Solutions of Calcium Lactate

C, %wt	26–40°C				40–60°C			
	α(orig), 10^{-4}	α(1), 10^{-4}	α(2), 10^{-4}	α(H_2O), 10^{-4}	α(исх), 10^{-4}	α(1), 10^{-4}	α(2), 10^{-4}	α(H_2O), 10^{-4}
0.82	3.30	3.11	3.06		4.64	4.49	4.88	
4.00	3.81	3.41	3.09	3.02	4.43	5.34	3.99	4.58
5.53	3.67	3.47	2.85		4.18	5.04	3.53	
7.71	3.14	3.50	2.66		4.75	4.51	3.02	

16.4 CONCLUSIONS

The structural-chemical transformations of calcium lactate (impure with calcium carbonate) initiated by the mechanical activation in a vortex mill have been studied.

A dependence of the structural transformations on treatment modes has been determined.

The mechanical activation is accompanied with chemical processes observed about $T \leq 200°C$: dehydration and partial decomposition of calcium lactate.

One mechanical activation cycle does not change the crystalline structure of the original powder; during the second cycle amorphization of the powder begins.

After 2 mechanical activation cycles, the structural state of the lactate anion changes it is indicated as additional lines in the X-ray diffractograms. The XPS analysis of the spectra evidences a change in the isomeric state of the lactate anion in the course of the mechanical activation.

ACKNOWLEDGMENTS

The study has been carried out within the Program of the RAS Presidium 12-P-2–1065.

KEYWORDS

- **Aqueous solutions**
- **Calcium lactate**
- **IR spectroscopy**
- **XPS**
- **XRD**

REFERENCES

1. Mashkovski, M. D. (2010). Lekarstvennye Sredstva, Izdanie Shestnadcatoe, Spravochnik, M. Novaya Volna, 1216p.
2. Lomovski, O. I. (2001). Prikladnaya Mechanokhimiya, Pharmacevtica I Medicinskaya Promyshlenost, Obrabotka Dispersed Materialov i Sred, *11*, 81–100.
3. Sakate, Y. et.al. (2005). Characterization of Dehydration and Hydration Behavior of Calcium Lactate and its Anhydrate, Colloids Surf, Biointerfaces, *46*, 135–141.
4. Gorbushina, A. I., Kanuinnikova, O. M., & Layanov, V. I. (2012). Vliyanie Mechanoaktivacii v Vihrevoy Melnuce na Structurnoe Sostoyanie I PhisicoHimicheskie Svoistva Lactate Calcia, 4.1 Tez Docl XXI Ross Molod Nauch Conf., 323–324.

CHAPTER 17

NONMONOTONIC CHANGE OF THE COMPOSITION OF SURFACE LAYERS OF THE MELT Co$_{57}$Ni$_{10}$Fe$_5$Si$_{11}$B$_{17}$ DURING ISOTHERMAL PROCESS

A. V. KHOLZAKOV and A. G. PONOMAREV

CONTENTS

17.1   INTRODUCTION

In the history of metal melt investigation there are roughly three stages [1]. At the first stage, it was established that a multi component metal melt was a much more complex object than an ideal solution and related to its design-theoretical constructs. At the second stage, it became clear that the poly terms of the properties of some liquid pure metals and especially alloys had inflections, jumps and other anomalies. The latter indicates that along with smooth structural changes there are jump-like changes, which are somewhat similar to phase transformations. The third stage of the experimental investigation of metal melts includes studying their properties during long isothermal holdings. The time dependences obtained have shown many different specific features, which, unfortunately, cannot be systemized and unambiguously explained yet.

As a rule, the structures of metal melts are examined for the formation of different formations in the melts such as micro groups, micro in homogeneities, clusters, etc. [2–5], and their bulk properties are investigated. However, the surfaces of metal melts have been studied very rarely due to the limited number of the experimental methods. Generally, they are techniques for measuring the surface tension of metal melts. In this chapter, the investigations have been performed using the method of X-ray photoelectron spectroscopy (XPS). The XPS method belongs to the methods for surface investigation. In 1986, in the Physico Technical Institute of the Ural Branch of the Russian Academy of Sciences, an X-ray electron magnetic spectrometer with the focusing of electrons by transverse magnetic field [6] was built for studying the liquid state of samples including metal melts.

In earlier works on binary and ternary systems [7–9], in the chemical composition of the surface layers of metal melts the changes have been found, which have non monotonic character. In addition some other authors have also found the changes in a number of properties, which have an oscillatory character in the process of non-equilibrium metals melt relaxation [10] and have offered different possible explanations of the results obtained.

17.2   EXPERIMENTAL PART

The X-ray photoelectron method was used for "in situ" studying the chemical structure and nearest surrounding of atoms in the liquid $Co_{57}Ni_{10}Fe_5Si_{11}B_{17}$ alloy within one experiment.

The study was carried out on the unique X-ray electron magnetic spectrometer designed for investigating both solid samples and their melts during a long period at different superheat temperatures of the melt (up to 1500 °C) [6]. The specific feature of the above spectrometer is the spatial separation of the investigation chamber and the energy-analyzer. Such device construction allows acting upon a sample, particularly the sample heating, without disturbing focusing properties of the energy-analyzer.

In the course of the experiment, the spectra of the core levels of all the elements present in the alloys were analyzed. At isothermal holdings, the ratio of the most intensive spectra was studied; in our case it was $CO_2p_{3/2}/Fe2p_{3/2}$.

The XPS method allows monitoring the surface state directly during the experiment. The presence of contaminants and the processes of the oxidation of the surface layers were controlled using C1 s- and O1 s-spectra directly in the course of the experiment.

Samples in the form of an amorphous ribbon were placed into the spectrometer chamber and heated. After the sample had been heated, the XPS measurements were performed in the liquid state within one experiment. During the experiment, the vacuum in the spectrometer chamber was 10^{-3} Pa. The measurements were conducted without disturbing the vacuum.

The transition from the crystalline state into the liquid state was controlled by a change in the oxygen content in the alloy surface layers. As the result of melting, the content of oxygen decreased almost to zero [12]. The experiment was conducted in the vacuum of 10^{-3} Pa. Such vacuum is not a super high vacuum; consequently, in such residual gas medium there are always the elements found in the air, only here they are in a much smaller amount. Since in the crystalline state, the metallic type of bond prevails, it is quite natural that the processes of the oxidation of elements take place on the surface. As the result of melting, the situation on the surface changes significantly. We believe that a cluster structure is formed, and the absence of oxygen in the surface layers indicates that the bond type has changed. While in the crystalline state there is the metallic type of bond, in the melt there is the covalent type of bond. Such bond type between the cluster elements is a necessary condition for the formation of clusters; otherwise any "open" bonds will be used by the oxygen of the residual gas medium for the formation of oxides in the surface layers of alloys. The above situation is not observed in the alloys under study.

Where the oxygen gets to during melting is still unclear nowadays. There are two possible variants:

1. it gets into the vacuum during melting, and
2. into the sample bulk.

Unfortunately, the method used does not allow studying the melt bulk characteristics, and the gas phase analysis with the use of a mass-spectrometer has not revealed any visible differences before and after melting.

17.3   RESULTS AND DISCUSSION

The study of the chemical structure of the surface layers composition of the Co-57Ni10Fe5Si11B17 alloys has been conducted in the liquid state at changes in temperature and isothermal holdings.

Before discussing the changes of the composition of the melt surface layers at isothermal holdings, let us consider the changes at heating. In the present work, the alloy transition into the liquid state and surface changes in the melt are of great interest.

Figure 17.1 shows a change in the composition of the Co57Ni10Fe5Si11B17 alloy surface layers at increasing temperature from 600° to 1500 °C [13].

In the crystalline state (up to 1050 °C) there are no any significant changes of the composition of the surface layers except the region of "pre-melting" which requires special consideration.

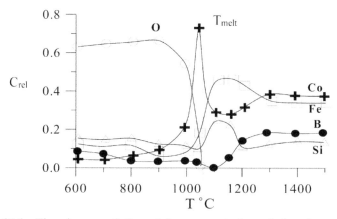

FIGURE 17.1 The changes of the relative concentration of the elements of the Co57Ni10Fe5Si11B17 alloy in the crystalline and liquid states [13].

The melting process of the Co57Ni10Fe5Si11B17 alloy is determined by the oxygen amount in the surface layers. A sharp decrease in the amount of oxygen corresponds to the sample melting (1050 °C) [12, 13].

In the liquid state, two temperature regions can be singled out. The first is from T_{melt} to 1200 °C, and the second from 1200 to 1500 °C.

In the first region, there are atoms Fe, Si and Co on the melt surface and in the second region the composition is supplemented by atoms B. Using the ides about the cluster structure of metal melts and the character of the change of the composition of the surface layers at increasing temperature, we can state that in the first temperature region there are mainly clusters based on Fe-Si, and in the second region, in addition to the Fe-Si clusters, the formation of Co-B-based clusters is possible. It should be noted, that the formation of more complex structural units on the melt surface is possible as well; however using only the XPS data, it is impossible to speak about any particular structural formations in metal liquids.

At the temperature of 1200 °C, the change of the surface layers composition takes place; in other words, the nearest surrounding of atoms changes. One type of cluster is replaced by the other one. The jump-like change of the surface layers composition associated with the change of the type of the short-range order can be interpreted as the structural transformation within the liquid state.

The results of studying the composition of the surface layers of the Co-57Ni10Fe5Si11B17 melt at isothermal holdings are of special interest. The ratio Co/Fe is chosen as a criterion. Figure 17.2 shows the changes of the Co/Fe ratio at different temperatures.

It is seen that at isothermal holdings the changes are of non monotonic character in the entire temperature range under study. To pass from the visual observation to the quantitative evaluation, a quantity named instability of surface layers is intro-

duced $\xi = \dfrac{1}{a}\sqrt{\dfrac{\sum(\bar{a}-a)^2}{N(N-1)}}$, where a is Co/Fe, N is the number of measurements at

each temperature. Figure 17.3 presents the change of x. It follows from Fig. 17.3 that the largest instability of the melt surface is observed in the region of the structural transformation when one type of clusters is replaced/ supplemented by the other type of clusters.

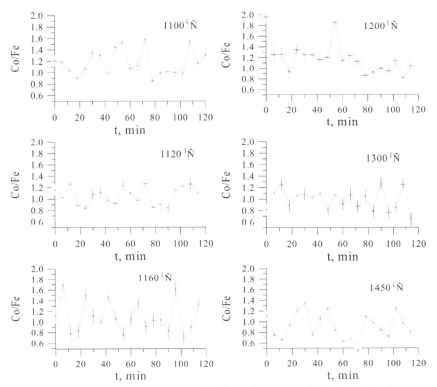

FIGURE 17.2   The change of the Co/Fe ratio in the surface layers of the Co57Ni10Fe5Si11B17 melt at isothermal holdings.

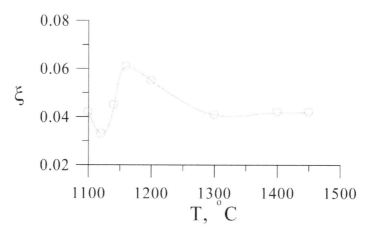

FIGURE 17.3 Instability of the surface layers of the Co57Ni10Fe5Si11B17 melt at increasing temperature.

For comparison, Fig. 17.4 shows the change of the *Co/Fe* ratio calculated with the use of the data in Fig. 17.1.

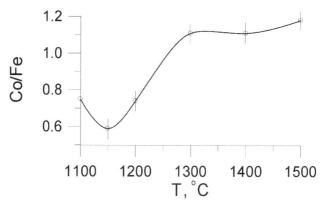

FIGURE 17.4 The change of the *Co/Fe* ratio at increasing temperature.

One can see that the curves in Figs 17.3 and 17.4 have a good correlation. The largest instability of the melt is observed in the first temperature region and close to the structural transition. At temperatures above 1200 °C the value of x is almost unchangeable. Such behavior of the Co57Ni10Fe5Si11B17 melt surface layers can be associated with the processes of the mixing of the clusters on the surface with the clusters from the near-surface region.

17.4 CONCLUSIONS

At changing temperature, in the Co57Ni10Fe5Si11B17 melt, the jump-like changes of the composition of the surface layers are found associated with the change of the type of the nearest surrounding of atoms, which are interpreted as structural transformations within the liquid state.

At isothermal holdings, the change of the composition of the surface layers is of non-monotonic character, which can be due to the redistribution of the cluster composition of the surface and the near-surface layers of the melt.

ACKNOWLEDGMENTS

The work is supported by grant 12-U-2-1001.

KEYWORDS

- **Chemical bond**
- **Clusters**
- **Metal melts**
- **X-ray photoelectron spectroscopy**

REFERENCES

1. Zamyatin, V. M., Baum, B. A., Mezenin, A. A., & Shmakova, Yu K. (2010). Time Dependences of Melts Properties, their Significance and Explanation Variants, Rasplavy, 5, 19–31.
2. Yelanski, G. N., & Yelanski, D. G. (2006). Structure and Properties of Metal melts, M Izd-vo Mosk, Gos, Vechernego Metallurg in-Ta, 228p.
3. Vatolin, H. A., Pastukhov, E. A., & Kern, E. M. (1974). Temperature Influences on the Structure of Melted Iron, Nickel, Palladium and Silicon, Doklady Akademii Nauka, 217(1), 127
4. Popel, P. S. (2005). Meta Stable Micro Heterogeneity of Melts in the Systems with Eutectic and Mono Tectic and its Influence on the Melt Structure After Hardening, Rasplavy, 1, 22–48.
5. Regel, A. R., & Glazov, V. M. (1980). Physical Properties of Electron Melts, M Nauka, 296p.
6. Trapeznikov, V. A., Shabanova, I. N., Varganov, D. V., et al. (1989). Creation of an Automated Electron Magnetic Spectrometer for Studying Melts, Otchyot VNTITs, 12880067297.
7. Kholzakov, A. V., Shabanova, I. N., & Ponomarev, A. G. (1999). XPS-Studies of Structural Transformation and Relaxation Processes in Transition Metal Melts, Elsevier Science Ltd, Vacuum, 53, 79–82.
8. Shabanova, I. N., Kholzakov, A. V., & Ponomarev, A. G. (2001). XPS-Studies of Structural Transformation and Relaxation Processes in Transition Metal Melts, Zr60Ti20Ni20, Ni72 Mo14B14, Ni81P19 and Ni82B18, Elsevier Science, B. V. J. Electron Spectroscopy and Related Phenomena, 114–116, 603–608.

9. Sapozhnikov, G. V., Kholzakov, A. V., Shabanova, I. N., & Ponomarev, A. G. (2004). Thermo-Structural Transformations and Instability of Ni-Based Melts According to the XPS Data, Poverkhnost, *10*, 63–66.

10. Baum, B. A., Igoshin, I. N., & Shulgin, D. B. (1988). On the Oscillatory Character of the Relaxation Process of Non-Equilibrium metal Melts, Rasplavy, *2(5)*, 102.

11. Siegbahn, K., Nordling, K., Fahlman, A., Nordberg, R., Hamrin, K., & Hedman, J. (1971). Electron Spectroscopy, M Mir, 493p.

12. Shabanova, I. N., Kholzakov, A. V., Chebotnikov, V. N., & Mukhina, Yu E. (1998). Cluster Structure of the Surface in Amorphous, Quasi-Crystalline and Liquid States of the $Zr_{60}Ni_{20}Ti_{20}$ Melt, Obshchestvo s Ogranichennoi Otvetstvennostyu, Metally, *5*, 106–109.

13. Kholzakov, A. V., Isupov, Yu N., & Ponomarev, A. G. (2011). Structural Transformations in Amorphous and Liquid States of the Fe-Co-Based Systems, Khimicheskaya Fizika i Mezoscopia, *13(4)*, 516–519.

CHAPTER 18

NICKEL(II) AND ETHYLENEDIAMINETETRAACETATE ACID INTERREACTION IN WATER SOLUTION OF AMINOPROPIONIC ACID

V. I. KORNEV and N. S. BULDAKOVA

CONTENTS

18.1   INTRODUCTION

α-aminopropionic acid (alanine, HAla) is known to generate complex compounds with different metals. As a result it is used as an effective Complexion agent. Alanine is generally absorbed by a liver. The synthesis speed of glucose from alanine and serine is much higher than synthesis from other amino acids. A liver and bowels absorb a lot of alanine [1, 2].

On the other hand, nickel (II) salts are used in various fields of science, technology and agriculture. Besides, nickel (II) is also supposed to be a biologically active metal. Too much nickel (II) in a body is said to cause toxic morphological change in a cell [3].

The combination of nickel (II) and pharmacologically active ligands (amino acids) bio effects as a part of coordination compounds causes toxicity reduction, on the one hand, and its biological activity increase as against its inorganic salts, on the other.

Besides, bio effect of nickel (II), that is a part of complex compound, depends on both metal concentrations, injected into the body, and coordination ligand character, as well as other factors.

The widespread usage and importance of nickel (II) salts and amino acids make an investigation of complex formation in these systems quite urgent. Investigation of polinuclear and hetaera ligand complexes' formation mechanism is essential for Coordination chemistry. Those cases when there are several ions of metals and some voluminous organic ligands in coordination sphere are of great interest. In this situation the problems of their compatibility and mutual influence in coordination sphere seem to be the most significant.

The formation of polinuclear complexes is connected with the fact that sometimes a polidentate ligand doesn't implement completely its valence in coordination sphere of one ion, which is a Complexion agent. The mononuclear complex can be a metal chelating ligand that is able to coordinate the second ion of metal of the same or some other nature. The formation of such complexes is established in triple systems Ni (II)–Edta–X, where X–Ox⁻ En, Gl⁻ et al. [4–6].

The object of this study is to define the models of complex formation, to calculate equilibrium constants and electronic-absorption spectra of complexes in system Ni (II)–HAla–Edta, that consists of Ni (II) and two ligands: the first is chelating (HAla), the second is at the same time chelating and bridged (EDTA). We haven't found any information about this topic in literature.

The information about processes in binary systems is essential for understanding of complex formation processes in triple systems. Particularly, the composition and complexes' stability in binary system Ni (II)–HAla under the same experimental conditions were studied.

18.2   EXPERIMENTAL RESEARCH

The study of the complexes formation was done with spectro photometric method. Spectrophotometer SF-2000 was used to expose the solution absorbency. Experiments were carried out at a device connected to quartz cuvette (1 cm). It allows titrating solutions. Such device allows to find out the solution absorbency, pH level and to control the solution temperature at the same time. The wave-length was set in the sphere 350–400 nm, 550–650 nm и 850–1000 nm to a precision of ±0, 1 nm. Absorbency dependence vs. pH, that describes the photolytic equilibrium, is observed best at the wave-length 985 nm. The algebraic dependences A=f(pH) was received with spectro photometric titration method. Bi distilled water was used as a comparing solution. Hydrogen ion activity was exposed with the equipment I-160 with the help of electrode ES-10601/7 and electrode ESR-10101. The equipment was calibrated with a standard buffer, made of fix anal, and tested with UPKP-1. Necessary pH level of the solutions was achieved with NaOH and HClO4 (analytical-reagent grade). NaClO4 (analytical-reagent grade) was used to have constant ionic strength $(I \approx 0, 1)$. It was $(20\pm2)^\circ$ C. Process solution of Nickel (II) per chlorate was made with dissolution of precise test portion of metal (analytical reagent quality) in per chloric acid. Complexion solution was made of fix anal. α-H Ala was prepared by means of preparation (extra pure) dissolution in distilled water. Programs CPESSP [7] and Hyp Spec [8] were used for mathematical treatment of the results. The complexes' models were made with program ACD/Labs [9].

18.3   RESULTS AND DISCUSSION

The basis of the complexes formation study in binary and triple systems was the change of absorption spectrum and the size of solution's absorbency of nickel (II) salt in the presence of a complexion and α -amino prop ionic acid.

Besides, theoretical models of a complex formation for triple systems (non registering hetero ligand ones) have been received. Some deflection connecting with processes of a complex formation of hetero ligand, homo and polinuclear complexes were revealed during the comparison of theoretical dependences of A=f (pH) with the experimental ones. By the form of the curves $A=f$ (pH) constructed for double and triple systems, we conclude that in all studied systems the complex formation is in a wide pH range.

Equilibrium study in the triple systems containing various hydroxide complexes and various proton forms of polidentat ligands is a challenge. If stability constants of complexes depend on the set of particles used for this system description, the choice of the correct set of complexes is of great importance. Fisher's test (finding out discrepancy between experimental and calculated data of absorbancy (A) for each system component) was used in the above programs to identify if it is necessary to take into account a complex form of a metal or a ligand. Programs allow to estimate parameters of equilibrium system, their stoichiometric, as well as to calculate the equilibrium constants of reactions and the stability constants of complexes. We applied the model of ion pairs for the description of chemical equilibriums in triple systems.

For complexes' identification we take into account three constants of mono metric hydrolysis of nickel (II) [10], four dissociation constants and two constants of a protonation of EDTA [11], and also two constants of dissociation and protonation α −HAla [10].

Some information about complex formation processes in binary systems Ni (II)– Edta and Ni (II)–HAla is necessary for equilibrium study in triple systems. Ni (II)– Edta system has been described in previous Ref. [12]. Some data about the second system are given in this article (Table 18.1).

TABLE 18.1   Some Information About Complex Compounds of Nickel (II) with α-Amino Propionic Acid at Ni(II): HAla = 1: 10, $I \approx 0,1$ (NaClO$_4$), $T = (20\pm2)°C$

Chemical formula	pH$_{opt.}$	Maximum accumulation share of complex, pH$_{opt}$, %	−lgK	lgβ
$Ni^{2+} + H_2Ala^+ \rightleftharpoons [NiAla]^+ + 2H^+$	5.8	60	6.49	5.39 ± 0.02
$Ni^{2+} + 2H_2Ala^+ \rightleftharpoons [NiAla_2] + 4H^+$	7.0	69	13.91	9.85 ± 0.08
$Ni^{2+} + 3H_2Ala^+ \rightleftharpoons [NiAla_3]^- + 6H^+$	10.0	100	22.78	12.86 ± 0.11

Equilibria modeling in triple systems was carried out by the analysis of electronic-absorption spectra and the experimental dependence of A=f (pH) for several wave-lengths according to the chosen model of a complex formation. In all cases A=f (λ) and A=f (pH) emulation has shown that models (non registering hetero ligand polinuclear complexes) in accordance with Fisher's test are in critical zone. We found out that in each system it is necessary to take into account such complexes formation. During the calculations we examined several models consisting of various sets of particles: [NiHiEdta]i–2 (i = 0–6), [Ni(OH)jEdta](j+2)–(j = 0–3), [NiAlan](2–n) (n = 1–3), [NimAlanEdta]2m-n–4 (m = 1–4, n = 1–8), and [Ni(OH)j]2–j (j = 0–3) and [HiEdta]i–4 (i = 0–6). However, Hydroxo complexes' portion here is minor, so that these particles' formation is not considered in the sequel.

The protolytic equilibrium, and also the complex structure in triple systems Ni (II)–HAla–Edta depend not only on acidity of the environment, but also from concentration of the components.

At a correlation of the components 1:1:1 in system Ni(II)–HAla–Edta (at pH > 7.0) there is an optical density decline of the curve of A=f(pH) as compared with correlation 1:0:1. The more alanine in this system, the bigger this effect is. It is clear that at pH > 7.0 we see hetero ligand nickel (II) complexion formation. As calculations showed, apart from mono ligand complexes of nickel (II) with EDTA, hetero ligand complex [NiAlaEdta]$^{3-}$ (α = 99%, pH = 10,0 at correlation 1:4:1) is generated in this system.

This formula reflects the complex formation:

$$[NiEdta]^{2-} + Ala^- \overset{K_1}{\rightleftharpoons} [NiAlaEdta]^{3-} \tag{1}$$

If there is twice as much nickel (II) and α-aminopropionic acid as EDTA in acid environment mononuclear and binuclear complex [(NiAla) 2Edta]$^{2-}$ are generated. The maximal accumulation share of the latter is 100% (pH$_{opt}$ = 9.0) (Fig. 18.1). The experimental and theoretical curves A=f (pH) are given in this and some other figures. Octahedral spheres of metals in a binuclear complex are bound among themselves with di-aminoetan bridges of a complexion. EDTA in this complex shows its maximum density, which is equal to 6.

For system Ni (II)-HAla-Edta (1:0:1 (*1*); 1:1:1 (*2*) and 1:4:1 (*3*);

Experimental curves (*1, 2, 3*), Ni^{2+} (*4*), [NiHEdta]$^-$ (*5*), [NiEdta]$^{2-}$ (*6*), [NiAlaEdta]$^-$ (*7*), C$_{Ni}$ = 2 × 10^{-2} mol/dm^3, λ = 985 nm

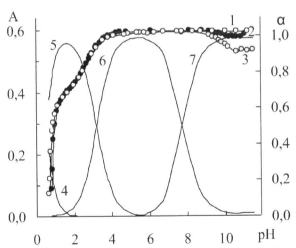

FIGURE 18.1 Dependence of optical density (A) and accumulation shares of complexes (α) vs. pH.

Below is a scheme of a binuclear hetaeraligand complex presented:

Alanine molecule has trans-configuration position to one of the glycine groups of EDTA, that is at outside position. The second glycine group has axial position. The vacant orbital of nickel (II) ion is kept by a water molecule.

This formula probably reflects the binuclear hetaeraligand complex formation in presence of alanine:

$$K_2$$

$$[NiEdta]^{2-} + [NiAla_2] \rightleftharpoons [(NiAla)_2Edta]^{2-} \qquad (2)$$

Experimental curve $A=f(pH)$ and complexes' fraction vs. pH at a correlation of the components Ni(II): HAla: Edta which is equal to 2:4:1 are given in Fig. 18.2b. It shows that in acid environment (up to pH = 4.0) there are much more mono-nuclear complexes [NiHEdta]$^-$ (α = 48% at pH = 2.0), [NiEdta]$^{2-}$ (α = 25% at pH = 3.5). At pH > 3.0 complex [(NiAla$_2$)$_2$Edta]$^{4-}$ is generated. Besides, there is complex [(NiAla)$_2$Edta]$^{2-}$ (α = 78% at pH = 4.5) in this solution.

The scheme presented here shows the transition of the first binuclear complex to the second one:

$$K_3$$

$$[(NiAla)_2Edta]^{2-} + 2Ala^- \rightleftharpoons [(NiAla_2)_2Edta]^{4-} \qquad (3)$$

Octahedral spheres of nickel (II) ions in a complex [(NiAla$_2$)$_2$Edta]$^{4-}$ are bound among themselves with di aminoetan bridge EDTA. Two metal carboxyl groups of different nitrogen atoms don't take part into coordination sphere formation.

Experimental curve (*1*), Ni^{2+} (*2*), [NiHEdta]$^-$ (*3*), [NiEdta]$^{2-}$ (*4*), [(NiAla)$_2$Edta]$^{2-}$ (*5*), [(NiAla$_2$)$_2$Edta]$^{4-}$ (*6*), C$_{Ni}$ = 2×10^{-2} mol/dm^3, λ = 985 nm.

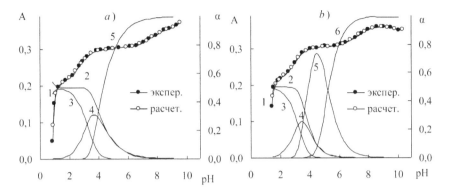

FIGURE 18.2   Dependence of optical density (A) and accumulation shares of complexes (α) vs. pH.

The analysis of electronic-absorption spectra of the solution, containing nickel (II) salt, HAla and EDTA (3:6:1), showed that in this system there is a hypos chromic shift of a d-d-absorption band. Mathematical modeling of equilibria in this system was done due to electronic absorption spectra examining for different wavelengths, as well as curves of A=f(pH) (Fig. 18.3a).

Complexes' fraction vs. pH indicates the fact that in acid environment (up to pH ≈ 3.0) mononuclear complexes of nickel with EDTA and free ions Ni^{2+} dominate. At pH > 3.0 complex [(NiAla)$_2$Edta]$^{2-}$ with maximum accumulation share 50% is emerged (at pH = 4.0). If you increase pH level, this complex decays according to Eq. (1). Complexion [(NiAla$_2$)$_3$ Edta]$^{4-}$ is formed (the scheme is presented).

$$K_4$$
$$[NiAla_2] + [(NiAla_2)_2\ Edta]^{4-} \rightleftharpoons [(NiAla_2)_3\ Edta]^{4-} \qquad (4)$$

The accumulation share of a three nuclear complex is 100% at pH = 9.0. The coordination spheres of two nickel (II) ions in three-nuclear complex are bound among themselves with Di Aminoetan Bridge of a complexion. Two covalent bonds of the EDTA metal carboxyl join the third ion of metal and a complex. Octahedral spheres of each metal contain two molecules of alanine.

Given below are the scheme (a) of a complex [(NiAla$_2$)$_3$Edta] $^{4-}$ and its 3D optimized model (b). Here atoms of ligands of EDTA and HAla, connected with nickel (II), are shown. Points are carbon atoms.

Modeling of complex processes in system Ni (II)–HAla–Edta at a correlation 4:8:1 showed that in the acid environment there are only mono complexes of nickel (II) EDTA. Two polinuclear hetaera ligand complexes are generated in the solution at pH > 3.0. In the sub acid environment complex [(NiAla) 2Edta]$^{2-}$ (α = 47%, pH = 4.6) is formed. If we consider neutral environment, complex [NiAla]$^+$ (α =

10% and pH = 5.9) is generated. While at pH > 5.0 in solution tetrahedral complex [(NiAla$_2$)$_4$Edta]$^{4-}$ is formed (maximum accumulation share 100% at pH > 8.0) (Fig. 18.3).

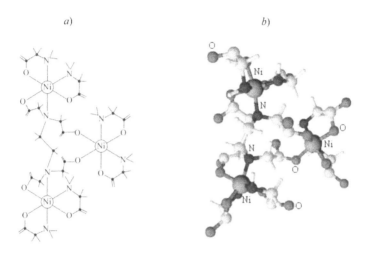

a)   *b)*

The formula below reflects the complex formation:

$$[NiAla_2] + [(NiAla_2)_3Edta]^{4-} \xrightleftharpoons{K_5} [(NiAla_2)_4Edta]^{4-} \tag{5}$$

In this complex octahedral spheres of four ions of metals are bound among themselves with EDTA molecule, which forms two covalent linkages with each of nickel (II) ions. Four bonds are formed at the expense of a de protonation of EDTA carboxyl groups. Four more coordination bonds emerge at the expense of two nitrogen atoms and two oxygen atoms of carboxyl groups. Coordination spheres of nickel contain two alanine molecules.

a)   for system Ni(II)-HAla-Edta (3:6:1);

b)   for system Ni(II)-HAla-Edta (4:8:1);

experimental curve (*1*), Ni^{2+} (*2*), [NiHEdta]$^-$ (*3*), [NiEdta]$^{2-}$ (*4*), [(NiAla)$_2$Edta]$^{2-}$ (*5*), [NiAla]$^+$ (*6*), [(NiAla$_2$)$_3$Edta]$^{4-}$ (*7*), [(NiAla$_2$)$_4$Edta]$^{4-}$ (*8*), $_{Ni}$ = 2×10^{-2} mol/dm^3, λ = 985 nm.

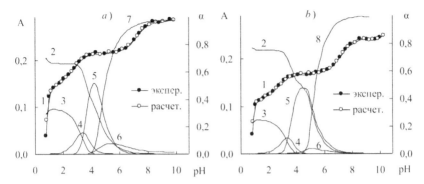

FIGURE 18.3 Dependence of optical density (A) and accumulation shares of complexes (α) vs. Ph.

TABLE 18.2 The Equilibria Constants of Reactions and Stability Constants of Nickel (II) Complexes with HAla and EDTA

Complex	Chemical formula number	lgK_i	$lg\beta$
$[NiAlaEdta]^{3-}$	(1)	2,98	$21,58 \pm 0,03$
$[(NiAla)_2Edta]^{2-}$	(2)	6,95	$35,70 \pm 0,06$
$[(NiAla_2)_2Edta]^{4-}$	(3)	13,78	$48,48 \pm 0,08$
$[(NiAla_2)_3Edta]^{4-}$	(4)	6,65	$64,98 \pm 0,04$
$[(NiAla_2)_4Edta]^{4-}$	(5)	2,77	$77,60 \pm 0,10$

This is tetra nuclear hetaera ligand Ni (II) complex's model:

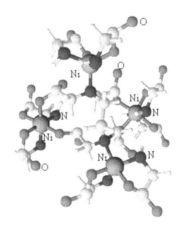

18.4   CONCLUSIONS

As a result, stability constants of nickel (II) complexes with EDTA and HAla are close to constants of complexes in system Ni (II)–En–Edta. The difference is caused by variety of molecule structure of di amino ethane and α-amino prop ionic acid. However, the general regularity of stability remains the same in complexes of the same composition. Nickel (II) is the core particle in polinuclear hetaera ligand complexes' formation, it determines this process. A polinuclear hetaera ligand complexes' composition depends not only on acidity of a solution and concentration ratios of components, but also on stereochemistry and ligands' density.

KEYWORDS

- **Alanine**
- **Complex**
- **Complexion**
- **EDTA**
- **Ligand**
- **Nickel(II)**
- **Spectro photometry**

REFERENCES

1. Komov, V. P., & Shvedova, V. I. (2008). *Biokhimiya* [Biochemistry], Moscow, Drofa, 640p.
2. Taylor, G. (1989). *Osnovy Organicheskoy Khimii* [Fundamentals of Organic Chemistry], Moscow, Mir, 384p.
3. Sadovnikova, L. K., Orlov, D. S., & Lozanovskaya, I. N. (2006). *Ekologiya I Okhrana Okruzhayuschey Sredy Pri Khimicheskom Zagryaznenii,* [Ecology and Environmental Protection During Chemical Pollution], Moscow, Vysshaya Shkola, 334p.
4. Barkhanova, N. N., Fridman, A. Y., & Dyatlova, N. M. (1972). *Obrazovanie Smeshannykh Soedineniy Perekhodnykh Metallov s Etilendiaminom I Etilendiamintetrauksusnoy Kislotoy* [Formation of Combined Compounds of Transition Metals with Ethylene Diamine and Ethylene Demine Tetra Acetic Acid], Journal of Inorganic Chemistry, *17(11)*, 2982–2988.
5. Barkhanova, N. N., Fridman, A. Y., & Dyatlova, N. M. (1973). Odno I Dvukhyadernye Soedineniya *Etilendiamintetraatsetatov Nikelya, Kobalta I Medias Ionami Glitsina I Oksalata v Rastvore* [Mono and Binuclear Compounds of Nickel, Cobalt and Copper Ethylene demine Tetra Acetate with Glycine and Oxalate Ions in Solution], Journal of Inorganic Chemistry, *18(4)*, 432–435.
6. Fridman, A. Y., Barkhanova, N. N., & Vshivtseva, O. E. (1981). *Obrazovanie Smeshannykh Poliyadernykh Kompleksnykh Soedineniy Nikelya (II) i Medi (II) s Etilen Diamine Tetraatsetat-Ionom i Bidentatnymi Ligandami v Rastvore* [Formation of Combined Poli Nuclear Com-

pounds of Nickel (II) and Copper (II) with Ethylene demine Tetra Acetate Ion and Bidentate Ligands in Solution], Journal of Inorganic Chemistry, *26(7),* 1783–1789p.

7. Salnikov, Y. I., Glebov, A. N., & Devyatov, F. V. (1989). *Poliy Adernye Kompleksy v Rastvorakh* [Poli Nuclear Complexes in Solutions] Kasan, Kasan University Publ., 288p.

8. ACD Labs, Available at: http://www. acdlabs.com (accessed 5 February 2011).

9. Lure, Y. Y. (1979). *Spravochnik Po Analiticheskoy Khimii* [Analytical Chemistry Reference-Book], Moscow, 480p.

10. Tereshin, G. S., & Tananaev, I. V. (1961). *Proizvedenie Rastvorimosti Etilendiamintetrauksusnoy kisloty* [Solubility Product of Ethylene demine tetra acetic acid] Journal of Analytical Chemistry, *16(5),* 523–526.

11. Kornev, V. I., & Buldakova, N. S. *Protoliticheskie i Koordinatsionnye Ravnovesiya v Vodnykh Rastvorakh Soley Nikelya (II), Etilen Diamin Tetra Uksusnoy Kisloty i Diaminoetana* [Protolytic and Coordination Equlibria in Water Solutions of Nickel(II) Salts, Ethylene Demine Tetra Acetic Acid and Di amino ethane], Chemical Physics and Mesoscopic Physics, *14(2),* 285–291.

CHAPTER 19

EFFECT OF MAGNETIC FIELD AND UV RADIATION ON STRUCTURAL STATE AND BIOLOGICAL PROPERTIES OF OLEIC ACID AND OLIVE OIL

V. I. KOZHEVNIKOV, O. M. KANUNNIKOVA, S. S. MIKHAILOVA, S. S. MAKAROV, and V. B. DEMENTYEV

CONTENTS

19.1   INTRODUCTION

Magnetic interaction physics supposes that diamagnetic molecules being put in a uniform static magnetic field tend to orient at right angle to the power lines, provided that these molecules have diamagnetic anisotropy. In case of a non-uniform constant magnetic field, there is a translation component that enhances the orienting effect of the field. All big molecules of organic compounds usually have such anisotropy. Therefore, the effect of constant magnetic field on the physical-chemical processes involving such anisotropic molecules is quite justified and predictable. This effect may be involved in phase-forming processes. Complex molecular systems are greatly influenced by both thermodynamic (dealing with phase-forming energy) and kinetic factors: due to the complexity and bulkiness of the structure units, macromolecules, they form a phase with partial Meta stable arrangement of the molecule fragments. Applying a constant magnetic field may provide an extra kinetic factor that determines diamagnetic orientation. Naturally, the share of Meta stable molecule positions during phase-formation is not prevailing.

This study is intended to analyze the effect of magnetic field and UV radiation on the structural state of olive oil almost completely comprised of triglycerides of oleic acid.

19.2   EXPERIMENTAL TECHNIQUE

The study objects were oleic acid and Extra Virgin olive oil mostly comprised of oleic acid triglycerides.

The density was measured with the use of pycnometer method, the accuracy was 0.2%. The capillary viscosity was measured with a viscometer ВПЖ2, with capillaries of 1.21 mm diameter. The rotation viscosity was measured with a DV-E device. The measurement accuracy amounted to 0.5 and 1.5% for capillary kinematic viscosity and rotation dynamic viscosity, respectively. The refraction indexes were measured with a refract meter IRF-454B2 M with an accuracy of ±0.0005.

19.3   RESULTS AND DISCUSSION

Rheological properties of olive oil (Table 19.1) display the values of olive oil rotation dynamic viscosity according to the spindle rotation time. In terms of rotation viscosity measurement, there are two distinct time intervals (0–1 h and over 4 h) when the rotation viscosity remains constant (within the experiment error); meanwhile, the viscosity in the first interval is more than that in the second interval. In 1 h after the last measurement, the rotation viscosity increases up to the initial value found by measuring viscosity after 0–1 h of spindle rotation.

In view of the actual regularity of the behavior of the rotation viscosity, olive oil may be characterized as a non-Newtonian fluid having both thyrotrophic (viscosity decrease) and rheopexic (viscosity increase) properties. Table 19.2 contains the rotation viscosity values measured for two temperatures, 18 h apart. The difference between the rotation viscosity values in Tables 19.1 and 19.2 is connected with different spindles used. The analysis of the results given in Table 19.2 demonstrates that the viscosity does not change at 25 °C (within the experiment error). Meanwhile, in measuring the rotation viscosity at 15 °C, the spindle rotation results in growing viscosity. It may be caused by the orientation of the triglyceride molecules during spindle rotation and formation of bigger molecular assemblies, that is, we see not just restoration of a disrupted structure but an aftereffect structuring of triglycerides.

TABLE 19.1 Dependence of Olive Oil Rotation Viscosity on Time and Speed of Spindle Rotation

Spindle rotation time, h	Spindle rotation speed, rpm	Rotation viscosity, mPa/sec	
		15°C	25°C
0	50	77	59.5
	100	–	59.4
0.5	50	78.7	58.1
	100	–	57.7
1.5	50	76.5	60.5
	100	–	59.3
4	50	73.9	51.2
	100	–	52.3
6	50	69.5	52.9
	100	–	52.7
6 + 1 h exposure	50	77.3	57.7
	100	–	58.2

TABLE 19.2 Change of Rotation Viscosity of Olive Oil in 18 h after Final Measurement

Initial viscosity measurement, mPa/s		Second viscosity measurement after 18 h, mPa/s	
15°C	25°C	15°C	25°C
100.8	71.7	107.1	69.9

The effect of treatment on the structure-sensitive properties of oleic acid and olive oil Treatment of oleic acid with UV radiation enhances the structure coherence, apparently, due to generation of intermolecular hydrogen bonds.

It is evidenced by the temperature coefficient of volume expansion decreasing after the exposure of the acid to UV radiation and increasing after the exposure to magnetic field (Table 19.3).

The magnetic field treatment has a destructing effect, probably, because of reorientation of the oleic acid molecules and destruction of some intermolecular hydrogen bonds of the initial acid. The effect of the magnetic field is more explicit than that of the UV radiation; therefore, the structure of the oleic acid after combined treatment is more coherent than that of the initial acid.

In case of olive oil, both kinds of treatment UV radiation and magnetic field reduce the volume expansion temperature coefficients, that is, increase the structure coherence. It should be noted that the oil exposed to large doses of UV radiation and magnetic field has anomalously high capillary viscosity at 15 °C. At 25 °C, the capillary viscosities of the treated and initial oils are slightly different (Tables 19.4 and 19.5). The measurement of the rotation viscosity also gives an anomalously high value at 15 °C for the oil exposed to a low dose of magnetic field.

Triglycerides of various compositions and structures produce main α-, β- and β'-forms of similar properties, since the polymorphism is mostly determined by acyl chains. A vitreous amorphous gamma-form may also be identified. The crystals of individual polymorphous forms of glycerides differ in the tilt angle of the hydrocarbon radicals of fatty acids with respect to the plane of the end groups. The stable beta-form of glycerides is the most high-melting. In it crystals, as in crystals of a beta-form, the hydrocarbon acid radicals are tilted at an angle of about 65 °C to the plane of the end groups. The radicals of the low-melting alpha-form are perpendicular to the plane of the end groups [1, 2].

TABLE 19.3   Effect of Treatment on the Structure-Sensitive Properties of Oleic Acid

Treatment	Density, g/cm3		Kinematic viscosity, mm²/s		Thermal coefficient of volume expansion, ×104, 1°C
	15°C	25°C	15°C	25°C	
Initial	0.8987	0.8928	45.51	30.57	6.5
UV, 5 min	0.8986	0.8946	44.48	29.88	4.4
UV, 50 min	0.8984	0.8934	45.41	30.57	5.6
MF, 5 min	0.8979	0.8908	45.37	30.42	7.9
MF, 50 min	0.8983	0.8909	45.27	30.48	8.2
MF+UV, 5 min	0.8984	0.8918	44.73	30.10	7.3
MF+UV, 50 min	0.8984	0.8918	45.94	30.84	7.3

TABLE 19.4 Effect of Treatment on the Refraction Index, Density and Thermal Coefficient of Volume Expansion of Olive Oil

Treatment	Refraction index	Density, g/cm^3		Thermal coefficient of volume expansion, ×104,
		15 °C	25 °C	1°C
Initial	1.4680	0.9189	0.9136	5.8
UV, 50 min	1.4685	0.9177	0.9134	4.7
MF, 5 min	–	0.9182	0.9131	5.6
MF, 50 min	1.4685	0.9179	0.9130	5.3
UV+MF, 50 min	1.4685	0.9175	0.9133	4.6

TABLE 19.5 Effect of Treatment on the Viscosity of Olive Oil

Treatment	Capillary viscosity, mm^2/s		Rotation viscosity, mPa/s	
	15°C	25°C	15 °C	25 °C
Initial	110.8	69.2	93.6	61.5
MF, 5 min	110.9	69.8	99.9	65.7
MF, 50 min	112.8	70.5	95.1	65.1
UV, 5 min	110.8	69.5	94.8	64.5
UV, 50 min	113.9	70.9	93.3	64.2
UV+MF, 5 min	108.8	68.6	90.0	63.9
UV+MF, 50 min	110.4	69.2	92.1	61.8

However, in the vast variety of the natural fats, the researchers keep finding new additional polymorphous forms; therefore, the classification of such forms is not established yet. The direction and products of the polymorphous transformations depend on the sample purity, presence and nature of crystallization inoculation, solvent, pressure, temperature, temperature change rate etc.

It may be presumed that, when in initial state, the form of oleic acid triglycerides dominating in the olive oil structure is the form having lower melting temperature and that the treatment forms a high-melting form of triglycerides. Therefore, the viscosity of the treated oil measured at 15 °C jumps, as the high-melting form begins to crystallize, thus slowing down the viscous flow. At 25 °C, the viscosity values for the initial and treated oils are just slightly different (less than 5%), since both forms are fluids.

The reduction of the thermal coefficient of volume expansion suggests multiplication of the hydrogen bonds. The treatment of olive oil with magnetic field results in reduction, even though minor one, of the thermal coefficient of volume expansion, that is, increase of the structure coherence. Probably, the treatment of olive oil influences both the intermolecular hydrogen bonds, just like in case of oleic acid, and the intra molecular hydrogen bonds. The UV radiation and magnetic field unequally promote the formation of intra molecular hydrogen bonds. Probably, the formation of the intra molecular hydrogen bond is also promoted by the generation of a high-melting structure of oleic acid triglycerides. In their turn, the intra molecular hydrogen bonds may contribute to the stability of the high-melting structural modification of triglycerides.

The differences in the structures of the initial and treated oils at temperatures close to 25 °C persist and manifest themselves in the difference in biological properties. Micro electrophoresis applied to buccal cells and blood cells was used to show the growth of the permeability of the cell membranes in the emulsion of the treated oils.

19.4   CONCLUSIONS

The study of the structure-sensitive properties of oleic acid and olive oil almost completely comprised of oleic acid triglycerides has concluded the following:

Olive oil behaves like a non-Newtonian fluid having both thixotropic (viscosity decrease) and rheopexic (viscosity increase) properties.

Treatment of oleic acid with UV radiation leads to higher structure coherence, apparently, caused by generation of intermolecular hydrogen bonds; magnetic field treatment has a destructuring effect, probably, caused by reorientation of the oleic acid molecules and destruction of some of the intermolecular hydrogen bonds of the initial acid; the effect of magnetic field is more explicit than that of UV radiation, so the structure of the oleic acid after combined treatment is more coherent than that of the initial acid.

The UV radiation and the magnetic field unequally promote the formation of intra molecular hydrogen bonds and a high-melting polymorphous modification of oleic acid triglycerides in the olive oil; the maximal coherence of the olive oil structure is recorded in case of combination of UV radiation and magnetic field.

ACKNOWLEDGMENTS

The authors thank A.A. Solovyov, Candidate of Medical Sciences, an assistant professor of the histology chair of Izhevsk State Medical Academy, for carrying out the micro electrophoresis survey.

The study has been performed within the Program of the RAS Presidium 12-P-2–1065.

KEYWORDS

- **Oleic acid**
- **Olive oil**
- **Polymorphous modifications**
- **Structure-sensitive properties**
- **Triglycerides**

REFERENCES

1. Vereschagin, A. G. (1971). Structurnyi Analis Prirodnyh Tryglyceridov, Uspehi Himii, XL (11), 1995–2028.
2. Ravich, G. B., & Curinov, G. G. (1952). Fazovaya Diagramma Trigliceridov (Prevrascheniya v Organicheskih Veschestvah v Tverdom Sostoyani, M Izd., AN SSSR, 158p.

CHAPTER 20

METASTABLE INTERSTITIAL PHASES BY BALL MILLING OF TITANIUM IN LIQUID HYDROCARBONS

A. N. LUBNIN, G. A. DOROFEEV, V. I. LADYANOV,
V. V. MUKHGALIN, O. M. KANUNNIKOVA, S. S. MIKHAILOVA,
and V. V. AKSENOVA

CONTENTS

20.1   INTRODUCTION

Substance treatment with the high-energy planetary ball mills is among the most efficient methods to produce strongly non-equilibrium states, such as supersaturated
solid solutions, nanocrystalline and amorphous phases etc. [1]. The deformation
defect generation rate for a solid being milled normally exceeds their thermal relaxation rate. Therefore, the stored defects density and related energy may reach critical
values enough to initiate non-equilibrium phase transformations. Consequently, the
mechano-chemical reaction mechanisms may greatly differ from the mechanisms
of conventional thermal synthesis [1]. A special class of mechano-chemical reactions is represented by the reactions accompanying mechanical milling (MM) of
metals in a liquid reaction medium, e.g. in liquid hydrocarbons (LHC). Titanium is
a carbide and hydride-forming element; therefore, study of its mechano-chemical
interaction with LHC's is of great interest. The literature data suggest that titanium
MM in LHC's may cause destruction of the LHC's and saturation of the metal with
carbon and hydrogen [2–5]. Most studies concern HCP→FCC phase transformation
in titanium, but they do not provide thorough analysis of the kinetics and structural
evolution mechanisms in the solid and liquid phases. There are no comparative studies addressing to detailed examination of titanium interacting with various liquid
hydrocarbons under high-energy ball milling, or to detection of the role of hydrocarbon chemical structure in this process.

   The goal of this study is comparative examination of the kinetics and mechanisms of the structural-phase transformations during mechanical milling of titanium
in liquid hydrocarbons (n-heptane, toluene) of various chemical structures and during subsequent low-temperature annealing.

20.2   EXPERIMENTAL TECHNIQUES

MM of titanium powder (99.9 wt %) was performed in a LHC medium in a planetary
ball mill AGO-2S with a carrier rotation speed of 890 rpm. The milling drums were
loaded with 200 g of balls of 8 mm diameter, 10 g of titanium and 50 mL of LHC. The
LHC's were represented by toluene ($C_6H_5CH_3$) and n-heptane (C_7H_{16}), 99.5 wt% pure.
The thermal stability of the phases obtained through MM was examined with 1 h annealing of samples at 550°C. The annealing was performed in two different media: in
argon at a pressure of 0.1 MPa and in dynamic vacuum at a residual pressure of 10^{-4}
Pa. The X-ray structural examination was carried out with a diffractometer D8 Advance in Cu K_α monochromated radiation. EVA software with PDF-2 phase bank was
used to carry out qualitative phase analysis. The X-ray diffraction patterns were processed with TOPAS 4.2 software using full-profile Rietveld method [6] to carry out
quantitative phase analysis, to determine the lattice parameters of the phases, to evaluate the crystallite volume-weight average size (D_v) and maximal lattice microstrains
(e). The instrumental line broadening was registered with a reference represented

by magnesium oxide prepared as per the method [7]. The probabilities of stacking faults (SF) in HCP phase were calculated from the diffraction patterns using Warren theory [8]. The surface composition and chemical bond types of the samples were analyzed by the X-ray photoelectron spectroscopy method using ES-2401 spectrometer. The spectra of the inner electron energy levels were excited with Mg K_α-radiation of 1253.6 eV. The mathematical treatment of the spectra was carried out according to the method [9]. The content of hydrogen, carbon and oxygen in the samples was found with gas-analyzer METAVAK-AK. The changing state of the LHC's during the mechano-chemical synthesis was surveyed through infrared spectroscopy (IR Fourier spectrometer FSM-1202), refractive index measurement (IRF-454 refractometer) and density measurements (Ostwald pycnometer).

20.3   RESULTS AND DISCUSSION

20.3.1   PHASE COMPOSITION

Figures 20.1 and 20.2 present the X-ray diffraction patterns of the samples after MM in toluene and n-heptane, respectively. They show that, as the MM time rises over 10 min, the X-ray diffraction patterns feature extra contributions (Figs. 20.1b and 20.2b) that can be regarded as formation of new phases, products of mechano-chemical interaction of Ti with LHC's. Figures 20.3 and 20.4 show the most meaningful pattern sectors after 30 min of MM resolved into components. Besides the broadened lines of α-Ti (HCP lattice), FCC reflexes and amorphous halo are observed. Figures 20.1c and 20.2c show decomposition of the diffraction patterns into components of different phases, as observed after 50 min MM. The analysis showed increase of parameters and ratio c/a for HCP lattice during MM in both media. It indicates formation of Ti-based HCP interstitial solid solutions. By the 50th minute of MM in n-heptane and toluene, the content of the FCC phase was 75 and 85 wt% respectively. The FCC lattice parameter was evaluated as $a = 0.4276_1$ nm for the sample obtained in toluene and $a = 0.4270_2$ nm for the sample obtained in n-heptane. The indexes by the numbers hereinafter indicate an error in the last significant digit. Besides the Ti-base phases, there is also 1–2 wt % of α-Fe-base phase impurity. The reason is the wear products of the milling bodies present in the sample. As is known through Ref. [10], FCC structure is typical for interstitial compounds of Ti with carbon and/or hydrogen. Our values of the FCC-phase parameter were significantly lower than the known data for hydride TiH_2 (0.4403 to 0.4445 nm [10]) and carbide TiC_x (0.43017 to 0.4328 nm [11]); meanwhile, they were close to the lowest values of the lattice parameters of titanium carbohydrides Ti (C, H) (0.4271 to 0.4301 nm [12]).

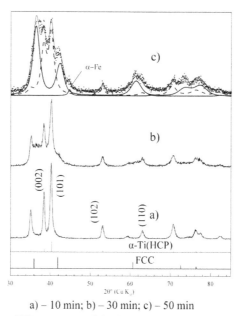

a) – 10 min; b) – 30 min; c) – 50 min

FIGURE 20.1   X-ray diffractograms of titanium powder after MM in toluene medium.

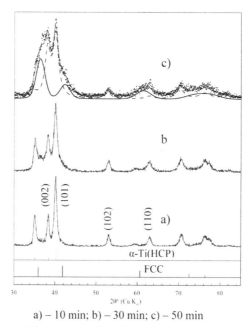

a) – 10 min; b) – 30 min; c) – 50 min

FIGURE 20.2   X-ray diffractograms of titanium powder after MM in n-heptane medium.

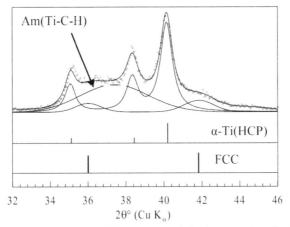

FIGURE 20.3   Segments of X-ray diffractograms of titanium powder after MM in toluene medium over 30 min.

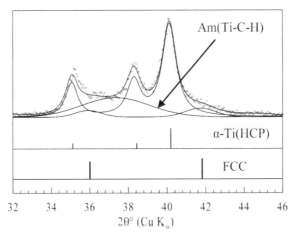

FIGURE 20.4   Segments of X-ray diffractograms of titanium powder after MM in n heptane medium over 30 min.

Titanium carbide TiC is a very thermo stable compound; the congruent melting temperature is $T_{melt} \approx 3100°C$ [10]. Titanium hydrides and carbohydrates decompose at 700 to 800°C [13]. The thermal stability of the obtained samples was studied at an annealing temperature below this interval. Figure 20.5 shows X-ray diffraction patterns of the powders after MM and subsequent 1 h annealing at 550 °C.

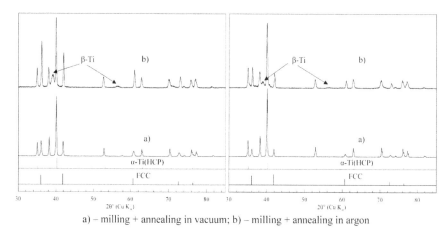

a) – milling + annealing in vacuum; b) – milling + annealing in argon

FIGURE 20.5   X-ray diffractograms of titanium powder after milling in toluene (left) or n-heptane (right) over 50 min and subsequent annealing (550 °C, 1 h).

We see that the annealing considerably decreases the intensity of the FCC phase reflexes. Thus partial reverse FCC→HCP transformation occurs. The values of the FCC-phase lattice parameters grow and reach 0.43051_2 and 0.43054_2 nm for vacuum annealing of the samples milled in toluene and n-heptane respectively. These parameters are within the parameter interval for stable titanium carbides TiC_x.

After argon annealing the lattice parameters are 0.42967_1 and 0.42951_1 nm and correspond with the values for stable titanium carbohydrates. The results of low-temperature annealing demonstrate that the FCC phase obtained through mechanical milling of titanium in LHC is Meta stable; most probably it is non stoichiometric titanium carbohydride. The diffraction patterns of the samples annealed in argon are special for reflexes of β-Ti, that is, a high-temperature form of Ti with a body centered cubic lattice. Apparently, upon heating up to 550 °C, β-phase was formed due to the presence of hydrogen that stabilized β-Ti by decreasing the formation temperature down to 300 °C (the temperature of eutectoid transformation in Ti-H system). Later, the formed phase was fixed with cooling.

20.3.2   CRYSTALLITE SIZES, LATTICE MICRO STRAINS AND STACKING FAULTS

After Ti MM in LHC, the size of α-Ti crystallites decreases down to 7–8 nm, while the lattice micro strains increase up to high values, 0.5%. The analysis of the profile and width of the sample MM diffraction lines demonstrated specific selective broadening of the reflexes with different hkl indexes. In particular the lines with $h-k = 3n+1$ ($n = 0, 1, 2, \ldots$) and $l \neq 0$ were broadened more than others. This selective broadening may be even seen in the diffraction patterns, (Figs 20.1a and 20.2a). According to Warren imperfect crystals diffraction theory [8, 14], this regular selective broadening of the reflexes resulted from stacking faults (SF) in the HCP-lattice. To find the stacking fault probability (or density), the physical broadening related to SF's only may be isolated from the general physical broadening which is known to be caused by blocks dispersion, lattice micro strains and SF's. B_{2q}, the value of SF-caused physical broadening, was determined with account of the fault affected lines, that is, the lines with $h-k = 3n+1$ ($n = 0, 1, 2, \ldots$) and $l \neq 0$. Meanwhile, the reference lines were the fault unaffected lines, that is, with indexes $h-k = 3n$ or $l = 0$, (002) and (110). The calculation of probabilities (fraction of faulty atomic planes) of deformation SP's (α) and growth SP's (β) dealt with the following combined equations [8]:

$B_{2\theta} = (360 / \pi)^2 \, l \, tg\theta \, (d / c)^2 \, (3\alpha + 3\beta)$	**For line** (102), l is even,
$B_{2\theta} = (360 / \pi)^2 \, l \, tg\theta \, (d / c)^2 \, (3\alpha + \beta)$	**For line** (101), l is odd,

where $B_{2\theta}$ is the physical width of the X-ray lines on the height half, caused by the presence of SF's, in degrees 2θ, d is the (hkl) inter planar spacing, c is the parameter of the HCP lattice.

SF probability calculation showed that prolonged MM results in greater quality of deformation SF's (Table 20.1). The probability of growth SF's β appeared negative; therefore, it proves that virtually there is no growth SF's. In the course of annealing the crystallites grew up to ~100 nm and the level of lattice micro strain was almost zero.

TABLE 20.1   Results of Probability Calculation for Deformation Stacking Faults (α) and Growth Stacking Faults (β) of Ti HCP after Milling in Liquid Hydrocarbons over 10 min

Milling medium	Indexes (hkl) and physical width $B_{2\theta}$ of X-ray lines, 2q deg.				A	β
	(002)	(101)	(102)	(110)		
Toluene	0	0.1353	0.1073	0	18 ′ 10^{-3}	−11 ′ 10^{-3}
n-Heptane	0	0.2256	0.0875	0	34 ′ 10^{-3}	−28 ′ 10^{-3}

20.3.3   X-RAY PHOTOELECTRON SPECTROSCOPY

Figures 20.6 and 20.7 shows C1 s, O1 s and Ti2p XPS spectra of the samples after 50 min MM in different LHC's and subsequent annealing at 550 °C in Ar medium and vacuum.

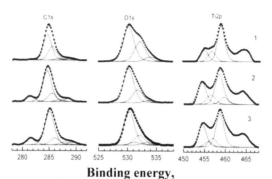

Binding energy,
1 – milling; 2 – subsequent argon annealing;
3 – subsequent vacuum annealing

FIGURE 20.6   C1 s, O1 s and Ti2p XPS-spectra of the samples milled in toluene, 50 m.

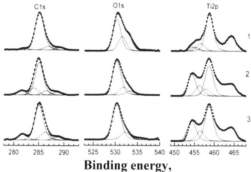

Binding energy,
1 – milling; 2 – subsequent argon annealing;
3 – subsequent vacuum annealing

FIGURE 20.7   C1 s, O1 s and Ti2p XPS-spectra of the samples milled in n-heptane, 50 min.

All the samples demonstrate C1 s spectral peaks with the binding energy E_b = 285 eV and more which correspond with the layer of adsorbed hydrocarbons and C-O fragments present on the particles surface. The lines on O1 s specter with E_b = 531–533 eV refer to the oxygen atoms of the adsorbed layer (organic compounds, water vapors, oxygen-containing gasses). Moreover, the analyzed surface particle layer includes a layer of titanium oxide TiO_2, as evidenced by a peak with E_b ~458.8 eV in Ti2p specter and a peal with E_b ~530.2 eV in O1 s specter. The spectra of powders after MM in LHC's may have lines with E_b ~282 eV in C1 s specter and with E_b~454.8 eV in Ti2p specter, evidencing presence of titanium carbide. A peak with E_{bond} ~456.6 eV is explained by chemical structures resembling oxi carbides Ti(C, O) appearing in the surface layer. Besides, a spectral line with E_b ~454.2 eV may be distinguished for the sample after MM in n-heptane. This line indicates presence of titanium carbohydrate. The element composition calculation showed that the relative titanium content in the surface particles layer after MM in n-heptane was 1.7 times greater than after MM in toluene. It proves that MM in toluene results in more adsorbed layers and surface compounds. Argon and vacuum annealing of the particles produced through titanium MM in LHC's results in disappearance of the titanium carbide line (E_b~454.8–455.0 eV) from Ti2p specter. Moreover, E_b of the line decreases in C1 s specter from titanium carbide to 281.2–281.6 eV. It proves that annealing in both environments causes titanium carbide to transform to titanium carbohydrate in the surface layer. Therefore, according to XPS, carbides are observed on the particle surfaces after MM, while carbohydrates are found in case of MM in n-heptane only, that is, in a liquid with greater hydrogen content. After annealing, the surface carbides transform to carbohydrates. This result shows that hydrogen as a reaction product concentrates mostly in the volume of powder particles, while carbon prevails on the surface. Annealing makes hydrogen leave the volume and its surface concentration grows, as a result the surface carbides transform to carbohydrates.'

20.3.4 CHEMICAL COMPOSITION

Chemical analysis (Table 20.2) shows that MM results in accumulation of carbon and hydrogen in the samples. The hydrogen-carbon ratio (H/C) is in agreement with that in LCH's themselves. Moreover, the samples contain an impurity of oxygen; it may be the result of the interaction with atmosphere during unloading from the mill. Vacuum annealing gives virtually complete hydrogen discharge, while argon annealing preserves hydrogen. Therefore, the findings in terms of carbon and hydrogen content prove the preliminary conclusion that the FCC-phase obtained during titanium MM in LHC is a non stoichiometric titanium carbohydrate, that is, with lack of interstitial atoms as compared with the stable titanium carbohydrates. Hydrogen thermal desorption during vacuum annealing is much more intensive than in argon atmosphere. Eventually, the vacuum annealing provides stable titanium carbide TiC and argon annealing provides stable titanium carbohydrate Ti(C, H).

TABLE 20.2 Results of Chemical Analysis of Interstitial Elements in the Samples After Milling and Annealing (H/C – Hydrogen-Carbon Atom Ratio in the Synthesis Products, in Toluene and n-heptane)

Time, milling medium and thermal treatment	Element content wt%/at%			H/C
	C	H	O	
50 min toluene	3.3/10.6	0.585/11.3	1.18/1.4	1.1
50 min toluene + annealing 550 °C in Ar	2.8/9.3	0.47/9.4	1.88/2.3	1.0
50 min toluene + annealing 550 °C in vacuum	2.37/8.7	0.036/0.8	1.8/2.5	0.09
50 min n-heptane	1.29/4.4	0.522/10.8	0.25/0.30	2.5
50 min n-heptane + annealing 550 °C in Ar	0.74/2.6	0.428/9.1	0.55/0.7	3.5
50 min n-heptane + annealing 550 °C in vacuum	0.68/2.6	0.02/0.5	1.05/1.5	0.15
Toluene C6H5CH3	91/47	9/53	-	1.1
N-heptane C7H16	84/30	16/70	-	2.3

20.3.5 IR-SPECTROSCOPY, DENSITOMETRY AND REFRACTOMETRY OF LIQUID HYDROCARBONS

IR-spectra of toluene and n-heptane taken after MM with titanium were also similar as compared with the initial LHC's, thus proving that 50-min MM causes no chemical transformation in LHC. Moreover, the density and refraction indexes remained constant and amounted to $d_{25}=0.8669$ g/ml, $n_D^{21}=1.4950$ and $d_{25}=0.6795$ g/ml, $n_D^{21}=1.3870$ for toluene and n-heptane respectively (the index stands for measurement temperature in °C). The obtained data show that, when in this MM phase, the reactions of toluene and n-heptane destruction are complete and produce final products C and H. The intermediate products neither accumulate in liquid phase, nor appear in the IR-spectra, nor change the density and refraction index.

20.3.6 MECHANISM OF MECHANO-CHEMICAL INTERACTION OF TITANIUM WITH LIQUID HYDROCARBONS

The obtained experiment data evidence titanium phase transformations caused by MM in LHC's. Production of titanium carbohydrates and amorphous base is participated by the interstitial elements (C, H) obtained through destruction of LHC's. Meanwhile, analysis (IR-spectroscopy and physical properties analysis) of the chemical structure of LCH molecules reveals no sign of LHC destruction, at least till 50 min of titanium MM. Hence, it may be concluded that all the transformations take place in the solid phase of titanium only. Proceeding from the obtained experimental data and present understanding of mechanical alloying [1], we may suggest the following mechanism of mechano-chemical interaction of titanium with LHC's. MM of titanium powder causes repetitive processes of fracture and cold welding of particles. In the presence of LHC's these processes are accompanied with adsorption of LHC molecules on the newly formed surfaces of the fractured Ti particles and accumulation of the LHC molecules inside of the powder particles with formation of the structures similar to laminar mechanical composites [1]. In other words, we observe mechano-activated absorption of hydrocarbon molecules by titanium powder. This structure favors mechano-chemical reactions due to the large contact surface between the chemicals. The LHC molecules embedded in metal particles may be hit and destructed by the balls that open their chemical bonds. The interaction of C and H atoms with Ti atoms is greatly influenced by the stored deformation defects of the metal structure: non-equilibrium nano-grain boundaries with elastic lattice micro distortions near the boundaries, deformation SF's, vacancies and interstitial atoms. The defects speed up the diffusion of the impurity atoms. The free C- and H-atoms produced through destruction of the LHC molecules penetrate along the grain boundaries of nano-crystalline titanium, segregate in the boundaries and distorted close-to-boundary zones, thus producing the amorphous phase Am(Ti-C-H) observed during the experiment. A Ti-base HCP interstitial solid solution is also formed. The main HCP → FCC transformation is present due to the accumulated deformation SF's and the C and H impurity atoms. Since ideal structures may have only two types of alternation of close-packed atomic planes, ABABAB in HCP and ABCABC in FCC, the SF in the HCP-crystal is a local area with FCC-stacking and they may act as nuclei of FCC-phase in HCP-matrix. Moreover, the presence of interstitial impurities simplifies generation of SF's [15]. The summarized mechano-chemical interaction of titanium with LHC's may be presented as follows:

$$Ti + C_y H_{y(liq)} \xrightarrow{\;1\;} Ti*(C_x H_y)_{absor} \xrightarrow{\;2\;} Am(Ti-C-H) + HCP - Ti(C,H) + FCC - Ti(C,H)_{metastab}$$

where $C_x H_{y(liq)}$ are hydrocarbons, toluene and n-heptane, $Ti*(C_x H_y)_{absorp}$ are titanium particles with LHC molecules absorbed in volume. The first stage starts with adsorption of LHC molecules on the surface of the powder particles and finishes with accumulation of the LHC molecules in the volume of particles in the manner described above (mechano activated absorption). The second stage supposes mechano-chemical interaction of titanium with absorbed hydrocarbon molecules accompanied with production of a number of solid substances. The transformations are stimulated by both enhanced diffusion of carbon and hydrogen in nano-crystalline titanium and generation of deformation SF's. The FCC-phase obtained through ball milling is meta-stable titanium carbohydrate with lack of interstitial atoms. Annealing of milled powders initiates relaxation of the stored defects and decomposition of the meta-stable phases accompanied by alternation of the chemical composition due to hydrogen thermo desorption.

20.4   CONCLUSIONS

The study of structural-phase transformations caused by ball milling of HCP titanium in the media of liquid hydrocarbons, toluene and n-heptane, has revealed the following:

1. Mechano-chemical interaction of α-Ti with toluene and n-heptane successively forms nanostructure in α-Ti, deformation-stacking faults in nanocrystalline α-Ti, amorphous phase Am (Ti-C-H) on the basis of titanium with carbon and hydrogen, interstitial solid solution α-Ti(C, H), meta stable FCC-phase. The hydrogen-carbon atom content ratios (H/C) in the products of mechano-chemical reaction correspond with those in the media of toluene (H/C=1.1) and n-heptane (H/C=2.3) used for synthesis.

2. FCC-phase is Meta stable titanium carbohydrate with lack of carbon and hydrogen; the phase originates and grows due to stacking faults in the HCP titanium lattice and saturation with carbon and hydrogen.

3. Low-temperature (550 °C) annealing of ball milled powders causes disintegration of meta stable titanium carbohydrates and alternation of the chemical composition. The main annealing products are stable phases: HCP Ti and titanium carbide TiC for vacuum annealing and HCP Ti (C, H), carbohydrate Ti(C, H) – for argon annealing. The difference in the annealing products is caused by higher rate of hydrogen thermo desorption in vacuum.

4. We have suggested a model of mechano-chemical interaction of the powder metal phase with liquid hydrocarbons; the model includes repetitive processes of failure, wetting and cold welding of the solid particles accompanied with adsorption of liquid molecules on the newly formed surfaces and their mechano activated absorption in the particles. The products of mechano activated destruction of the absorbed molecules, carbon and hydrogen atoms, saturate the imperfect nano-crystal as per the mechanisms of enhanced diffusion with new phases formed.

ACKNOWLEDGEMENTS

The authors thank V.V. Strelkov for performing the chemical analysis. The study has been funded by the programs of the Presidium of Ural Branch of Russian Academy of Science (Project 12-T-2-1014).

KEYWORDS

- **Interstitial phases**
- **Liquid hydrocarbons**
- **Mechano-chemical reactions**
- **Titanium**

REFERENCES

1. Suryanarayana, C. (2004). Mechanical Alloying and Milling, New York, marcel Dekker Inc., 466p.
2. Manna, I., Chattopadhyay, P. P., Nandi, P., Banhart, F., & Fecht, H. J. (2003). Formation of Face-Centered-Cubic Titanium by Mechanical Attrition, J. Appl., Physics, *93(3)*, 1520–1524.
3. Seelam, M. R., Barkhordarian, G., & Suryanarayana, C. (2009). Is There a Hexagonal-Close-Packed (HCP) → Face-Centered-Cubic (FCC) Allotropic Transformation in Mechanically Milled group IVB Elements? J. Mater Research, *24(11)*, 3454–3461.
4. Suzuki, T. S., & Nagumo, M. (1995). Metastable Intermediate Phase Formation at Reaction Milling of Titanium and n-Heptane, Scripta Metallurg, Mater, *32(8)*, 1215–1220.
5. Phasha, M. J., Bolokang, A. S., & Ngoepe, P. E. (2010). Solid-State Transformation in Nano-crystalline Ti Induced by Ball Milling, Mater Letters, *64*, 1215–1218.
6. Young, R. A. (1993). Introduction to the Rietveld Method, Oxford, Oxford University Press, 299p.
7. Pratapa, S., O'Connor, B. (2001). Development of MgO Ceramic Standards for X-Ray and Neutron Line Broadening Assessments, Adv., X-ray Analysis, *45*, 41–47.
8. Warren, B. E. (1990). X-Ray Diffraction, N.Y. Dover Publ., 381p.
9. Povstugar, V. I., Shakov, A. A., Mikhailova, S. S., Voronina, E. V., & Elsukov, E. P. (1998). Resolution of Complex X-ray Photoelectron Spectra Using Fast Discrete Fourier Transformation with Improved Convergence Procedure: Assessment of the Usability of the Procedure, J. Anal. Chem., *53(8)*, P. 697–700.
10. Goldschmidt, H. J. (1967). Interstitial Alloys, Plenum, New York, Butter Worths, London.
11. Zueva, L. V., & Gusev, A. I. (1999). Effect of Non Stoichiometric and Ordering on the Period of the Basic Structure of Cubic Titanium Carbide, Physics Sol., State, *41*, 1032–1141.
12. Khidirov, I., Mirzaev, B. B., Mukhtarova, N. N., Kholmedov, Kh M., Zaginaichenko, Yu B., & Schur, D. V. (2008). Neutron Diffraction Invastigation of Hexagonal and Cubic Phases of Ti-C-H System, Carbon Nanomaterials in Clean Energy Hydrogen Systems, NATO Science for Peace and Security Series C, Environmental Security, 663–678.
13. Dolukhanyan, S. K. (2005). SVS-method of Producing of Hydrogen Storage Units, International Scientific Journal for Alternative Energy and Ecology (ISJAEE), *11*, 13–16 (in Russian).
14. Fadeeva, V. I., Leonov, A. V., Szewczak, E., & Matyja, H. (1998). Structural Defects and Thermal Stability of Ti(Al) Solid Solution Obtained by Mechanical Alloying, Mater Sci. Engin. A, *242*, 230–234.
15. Hirth, J. P., & Lothe, J. (1968). Theory of Dislocations, New York, McGraw-Hill.

CHAPTER 21

PARAMETERS OF MOBILITY FOR POLLUTION BY SODIUM ARSENITE FOR SOIL OF KAMBARKA DISTRICT

V. G. PETROV, M. A. SHUMILOVA, and O. S. NABOKOVA

CONTENTS

21.1   INTRODUCTION

In laboratory tests we have found that the sodium arsenite has high mobility in the upper contaminated layer of soil in the simulation of atmospheric effects such as rain precipitation [1, 2]. This is primarily due to the fact that the sodium arsenite is a highly soluble compound. In addition, arsenic in sodium arsenite is in anionic form, which also affects the ion exchange processes in soils. We have shown that cationic forms of metal are kept better in soil [3, 4]. Anthropogenic impact of sodium arsenite on the soil could be in the works for the destruction of lewisite in Kambarka Udmurt Republic in 2006–2009, as it is a product of the decomposition of this toxicant by alkaline hydrolysis [5, 6]. Therefore, investigation related to the definition of the mobility parameters of sodium arsenite in soils Kambarka district is of interest. These data can be used both in the development of environmental monitoring and sanitation of contaminated areas [7].

21.2   METHODS AND MATERIALS

Different types of soil Kambarka district for research were provided by the Regional Centre on control and monitoring in the Udmurt Republic [8]. Sodium arsenite – $NaAsO_2$, mark "high-purity," was obtained from the Karaganda Institute of Chemical Technology (Republic of Kazakhstan). Study mobility of sodium arsenite in contaminated soil samples was carried out in a special laboratory stand simulating the effect of atmospheric precipitation in the form of rain [9]. Contamination by sodium arsenite was performed in an amounts of 12.5 and 25 MPC (MPC of As = 2.0 mg/kg air-dry soil) [10]. Determination of arsenic in solution was determined by atomic absorption spectrophotometry on the device "Shimadzu" – AA7000. Humus content in the soil samples were performed by method of Tyurin [11]. To determine the parameters of the pollutant mobility in his separation from the soil into the water under the influence of atmospheric precipitation in the form of rain, we have used the method of calculating the kinetics of this process, specifically designed for experimental stand [1, 9].

21.3   RESULTS AND DISCUSSION

Table 21.1 shows the humus content and level density of dry soil samples. Soils can be characterized as sand with different humus content from 1 to 3 wt%. Tables 21.2 and 21.3, and Figs. 21.1–21.3 show how the degree of separation of sodium arsenite from contaminated soil samples depend on the water volume passed through the soil.

TABLE 21.1 Characteristics of the Studied Soil Samples from Kambarka District Udmurt Republic

Number of soil samples	Humus, wt%	Level density, kg·m^{-3}
№ 1	1.00	1312
№ 2	2.95	1147
№ 3	1.55	1348
№ 4	2.22	1158
№ 5	1.73	1177

TABLE 21.2 The Degree of Separation Arsenic by Passing Water Through the Contaminated Soil. The Fraction Volume is 50 mL, the Rate of Water Filtration Through Contaminated Sample $\omega \sim 6.94 \times 10^{-3}$ mL·s^{-1}, Pollution by Sodium Arsenite 12.5 MPC for the As

№	The volume of water passed through the sample, mL	The degree of separation-α, the share of initial		
		Soil sample № 1	Soil sample № 4	Soil sample № 5
1	50	3.732×10^{-2}	6.873×10^{-2}	3.560×10^{-3}
2	100	1.069×10^{-1}	1.459×10^{-1}	7.762×10^{-3}
3	150	1.891×10^{-1}	2.579×10^{-1}	5.589×10^{-2}
4	200	2.884×10^{-1}	2.948×10^{-1}	1.110×10^{-1}
5	250	3.739×10^{-1}	3.257×10^{-1}	2.061×10^{-1}
6	300	4.241×10^{-1}	3.444×10^{-1}	3.090×10^{-1}
7	350	4.812×10^{-1}	3.633×10^{-1}	3.928×10^{-1}
8	400	5.451×10^{-1}	3.799×10^{-1}	4.703×10^{-1}
9	450	5.968×10^{-1}	4.014×10^{-1}	5.395×10^{-1}
10	500	6.422×10^{-1}	4.251×10^{-1}	6.072×10^{-1}
11	550	6.752×10^{-1}	4.354×10^{-1}	6.506×10^{-1}
12	600	7.097×10^{-1}	4.412×10^{-1}	6.886×10^{-1}
13	650	7.511×10^{-1}	4.487×10^{-1}	7.242×10^{-1}
14	700	8.051×10^{-1}	4.555×10^{-1}	7.857×10^{-1}
15	750	8.141×10^{-1}	4.592×10^{-1}	8.037×10^{-1}
16	800	8.185×10^{-1}	4.621×10^{-1}	8.262×10^{-1}

Calculation of constants the rate of separation contaminant from contaminated soil for experimental stand is shown in Refs. [1, 9]. It can be shown that the order of the kinetic equation is determined from the equation:

$$\frac{V_1}{V_2}\frac{\omega_2}{\omega_1} = \left(\frac{C_{0,2}}{C_{0,1}}\right)^{n-1},$$

(1)

where V_1, V_2 – water volume passed through the soil samples with different initial concentration of pollutant $C_{0,1}$, $C_{0,2}$, in which will be allocated the same amount of matter as a proportion of original content – α; at a rate of water filtration – ω_1, ω_2.

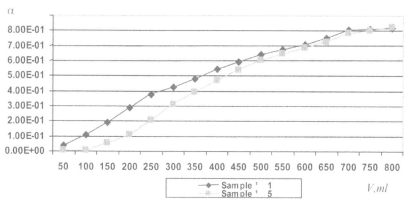

FIGURE 21.1   The dependence of separation degree α (as a fraction of the initial content) $NaAsO_2$ from water volume V (mL), passed through the soil samples № 1, 5. Contamination of sodium arsenite 12, 5 MPC of As.

TABLE 21.3   The Degree of Separation Arsenic by Passing Water Through the Contaminated Soil. The Fraction Volume is 50 mL, the Rate of Water Filtration Through Contaminated Sample $\omega \sim 6.94 \times 10^{-3}$ mL·s^{-1}, Pollution by Sodium Arsenite 25 MPC for the As

№	The volume of water passed through the sample, mL	The degree of separation-α, the share of initial		
		Soil sample № 2	Soil sample № 3	Soil sample № 4
1	50	5.036×10^{-2}		4.479×10^{-2}
			3.852×10^{-2}	
2	100	1.046×10^{-1}		1.080×10^{-1}
			7.853×10^{-2}	
3	150	2.671×10^{-1}		1.693×10^{-2}
			2.086×10^{-1}	
4	200	4.378×10^{-1}	3.362×10^{-1}	2.342×10^{-1}
5	250	5.374×10^{-1}	4.458×10^{-1}	3.040×10^{-1}

TABLE 21.3 *(Continued)*

6	300	6.210×10^{-1}	5.611×10^{-1}	3.780×10^{-1}
7	350	6.622×10^{-1}	6.242×10^{-1}	4.625×10^{-1}
8	400	6.756×10^{-1}	6.632×10^{-1}	5.650×10^{-1}
9	450	6.879×10^{-1}	6.978×10^{-1}	6.132×10^{-1}
10	500	7.077×10^{-1}	7.377×10^{-1}	6.393×10^{-1}
11	550	7.225×10^{-1}	7.800×10^{-1}	6.665×10^{-1}
12	600	7.336×10^{-1}	8.217×10^{-1}	6.898×10^{-1}
13	650	7.381×10^{-1}	8.631×10^{-1}	7.180×10^{-1}
14	700	7.416×10^{-1}	9.042×10^{-1}	7.507×10^{-1}
15	750	7.446×10^{-1}	9.358×10^{-1}	7.738×10^{-1}
16	800	7.462×10^{-1}	9.657×10^{-1}	7.823×10^{-1}

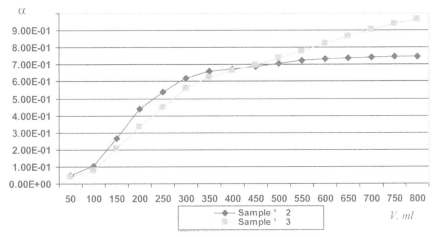

FIGURE 21.2 The dependence of separation degree α (as a fraction of the initial content) $NaAsO_2$ from water volume-V (mL), passed through the soil samples № 2, 3, Contamination of sodium arsenite 25 MPC of As.

In [1] it was shown that the order of separation sodium arsenite from contaminated soil differs from the heavy metal oxides and close to 2. For soil sample number 4 was also made calculation of order kinetic equation based on the experimental data given in Tables 21.2, 21.3 and depicted in Fig. 21.3. Table 21.4 shows the results of calculation.

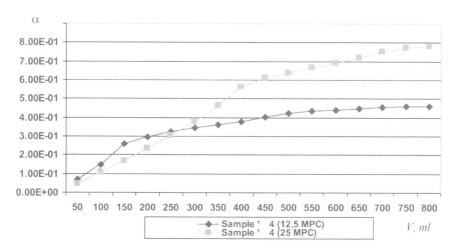

FIGURE 21.3   The dependence of separation degree α (as a fraction of the initial content) NaAsO$_2$ from water volume V (mL), passed through the soil sample № 4 with contamination of sodium arsenite 12.5 and 25 MPC of As.

TABLE 21.4   Values of the Order of Separation NaAsO$_2$ from Soil Sample Number 4, Defined on the Basis of experimental data given in Table 21.3 and illustrated in Fig. 21.3

The degree of separation-α, the share of initial	The order of separation-n
0.37	1.40
0.40	1.52
0.43	1.80
Mean n	1.57

From Table 21.4 shows that the order of separation NaAsO$_2$, as in Ref. [1], is close to 2. Accordingly, the observable rate constant for separation can be determined from the formula:

$$\kappa_H = \frac{\omega}{V}\left(\frac{\alpha}{1-\alpha}\right)$$

(2)

Using data from Tables 21.2, 21.3, we can calculate the rate constants for the separation NaAsO$_2$. Table 21.5 shows the results of calculation of the rate constants for the separation NaAsO$_2$ from contaminated soil samples.

Table 21.5 shows that there is a dependence of the rate constant for separation NaAsO$_2$ from humus content in the soil. Increase the humus content in the soil reduces the rate constant for separation of sodium arsenite under identical experimental conditions (see Fig. 21.4). This conclusion is also supported by the results of Ref. [1].

TABLE 21.5 Values of Observable Rate Constant for Separation $NaAsO_2$ for different Studied Soil Samples under Experimental Conditions

Soil sample	Pollution, MPC of As	κ_n, c^{-1}
№ 1	12.5	$3,912 \times 10^{-5}$
№ 2	25	$2,551 \times 10^{-5}$
№ 3	25	$2,442 \times 10^{-4}$
№ 4	12.5	$7,453 \times 10^{-6}$
№ 4	25	$3,117 \times 10^{-5}$
№ 5	12.5	$4,124 \times 10^{-5}$

Calculation the effective half-life $(T_{0.5})$ of pollutant when exposed to atmospheric rainfall specific for the region shows that sodium arsenite at the investigated levels of pollution has a high mobility in soils of Kambarka district. $T_{0.5}$ is a few days and is close to the values obtained for sand in Ref. [1].

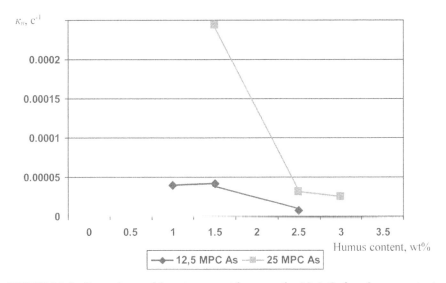

FIGURE 21.4 Dependence of the rate constant for separation $NaAsO_2$ from humus content.

21.4 CONCLUSIONS

Sodium arsenite has high mobility in soils of Kambarka district under the influence of atmospheric precipitation in the form of rain. Mobility parameters for different soil samples are close to the values obtained previously for the sand. This should be considered when sanitation of contaminated areas and monitoring the impact of pollutant. Technical measures for sanitation of contaminated areas can be developed relating to the territory of the district Kambarka district on sand soil samples. Increase the humus content in the soil reduces the rate of separation sodium arsenite from contaminated soils.

KEYWORDS

- **Mobility parameters**
- **Sodium arsenite**
- **Soil contamination**

REFERENCES

1. Shumilova, M. A., Petrov, V. G., & Nabokova, O. S. (2012). Kinetics of Separation Sodium Arsenite from Contaminated Soil, Chemical Physics and Mesoscopy, *14(4)*, 626–632.
2. Petrov, V. G., Shumilova, M. A., Nabokova, O. S., & Sergeev, A. A. (2012). Valuation Mobility in Soil Sodium Arsenite Pollution, Bulletin of Udmurt University, Series, Physics, Chemistry, *1*, 98–104.
3. Petrov, V. G., Shumilova, M. A., Charaldina, E. A., & Esenkulova, S. V. (2012). Comparison of Mobility in Soil Chromium Compounds, Bulletin of Udmurt University, Series, Physics, Chemistry, *2*, 69–73.
4. Petrov, V. G., Shumilova, M. A., Lopatina, M. V., & Aleksandrov, V. A. (2012). Study of the Sorption of Copper (2+) in the Soil, Bulletin of Udmurt University, Series, Physics, Chemistry, *2*, 74–77.
5. Umyarov, I. A., Kuznetsov, B. A., Krotovich, I. N. et al. (1993). Destruction and Disposal of Lewisite and Mustard, Russian Chemical Journal, *37(3)*, 25–29.
6. Petrunin, V. A., Baranov, U. I., Kuznetsov, B. A. et al. (1995). Mathematical Modeling of the Alkaline Hydrolysis of Lewisite, Russian Chemical Journal, *39(4)*, 15–17.
7. Nabokova, O. S. (2012). Perfection of Environmental Monitoring of Soils at Technogenic Impact of Arsenic Compounds for Facilities for the Destruction of Chemical Weapons, Bulletin of Udmurt University, Series, Physics, Chemistry, *4*, 59–62.
8. Shumilova, M. A., Nabokova, O. S., & Petrov, V. G. (2011). Features of the Behavior of Man-Made Arsenic in Natural Objects, Chemical Physics and Mesoscopy, *13(2)*, 262–269.
9. Petrov, V. G., & Shumilova, M. A. (2012). Way to Study in the Laboratory Mobility Techno Genic Pollution in Soil, Chemical Physics and Mesoscopy, *14(2)*, 257–260.
10. Bandman, A. L. et al. (1989). Harmful Chemicals. Inorganic Compounds V–VIII Groups, A Reference Book, Chemistry, 592p.
11. Mineev, V. G. Ed. (2001). Practical on Agricultural Chemistry, Publishing House, M "The Moscow State University," 689p.

CHAPTER 22

X-RAY PHOTOELECTRON STUDY OF THE INFLUENCE OF THE AMOUNT OF CARBON NICKEL-CONTAINING NANOSTRUCTURES ON THE DEGREE OF THE POLYMETHYLMETHACRYLATE MODIFICATION

I. N. SHABANOVA, V. I. KODOLOV, N. S. TEREBOVA,
V. A. TRAPEZNIKOV, N. V. LOMOVA, G. V. SAPOZHNIKOV,
A. V. OBUKHOV, YA. A. POLYOTOV, and N. YU. ISUPOV

CONTENTS

22.1   INTRODUCTION

One of the main existing hypotheses about the nanomaterial formation is based on the interatomic interaction of input components and their amount. It means that one of the main tasks is the development of diagnostic methods, which will allow to controls intermediate and final results in the creation of new materials. The main role in the diagnostics of materials belongs to X-ray methods. In this connection, the development of the method of X-ray photoelectron spectroscopy (XPS) becomes urgent indeed [1].

The goal of the work is the XPS study of the influence of the content of carbon nickel-containing nanostructures (C/Ni nanostructures) on the degree of the poly-methylmethacrylate (PMMA) modification.

22.2   EXPERIMENT PART

Carbon metal-containing nanostructures are nano-films from carbon fibers associated with metal-containing clusters, which are formed in nano-reactors of polymer matrices. Carbon metal-containing structures were prepared by a unique method in the conditions of low temperatures (not more than 400°C), which is a pioneer method [2].

To provide a uniform distribution of carbon metal-containing nanostructures throughout the bulk of a modified polymer is a very difficult task. A conventional method for the uniform distribution of a nano-addition in the bulk of a material is the obtaining of nanoparticle fine-dispersed suspensions in different media.

Methylene chloride was used for dissolving PMMA. The method for modifying PMMA with C/Ni nanostructures included the preparation of a fine-dispersed suspension (FDS) containing C/Ni nanostructures. For the preparation of the FDS, a polymer solution was prepared in methylene chloride; after that, the C/Ni nanostructures and the prepared solution were mixed. For the uniform distribution of the C/Ni nanostructures in the polymer solution, the above mixture was treated by ultrasound. Then, the films were prepared by evaporating the solvent (at the temperature increase up to 100 °C).

The XPS investigations were conducted on an X-ray electron magnetic spectrometer with the device resolution of 10^{-4} and luminosity of 0.085% at the excitation by AlKα-line at 1486.5 eV in vacuum of 10^{-8}–10^{-10} Pa. A magnetic spectrometer has a number of advantages connected with constructive capabilities of X-ray electron magnetic spectrometers in comparison with electrostatic ones; they are the consistency of luminosity and resolution regardless of electron energy, high contrast of spectra, and the possibilities for external actions upon a sample during measurements [1].

The investigations of the change of the polymer structure during nano-modification were based on studying the change of the shape of a C1s-spectrum.

A method for the identification of C1s-spectra was developed based on the use of their satellite structure, which allowed determining the chemical bond of elements, the nearest surrounding of atoms and the type of sp-hybridization of carbon valence electrons in the nanostructures and the materials nano-modified with them [3, 4]. In the C1s-spectrum of single-layer and multilayer carbon nanotubes, there were two satellites at 306 and 313 eV characteristic of C-C bonds with the sp^2 and sp^3–hybridization of valence electrons on carbon atoms [3]. The change of the structure of the PMMA modified with carbon nickel-containing nanostructures was studied. The content of the nanostructures in the studied PMMA samples was 10^{-5}–3%. PMMA without nano-additions was used as a reference sample.

22.3   RESULTS AND DISCUSSION

Figure 22.1 shows the C1s-spectrum of a carbon nickel-containing nanostructure consisting of three components Ni-C –283 eV, C-H –285 eV, and C-C (sp^3) –286.2 eV. The presence of a little amount of the C-H-component indicates incomplete synthesis of nanostructures from the polymer matrix.

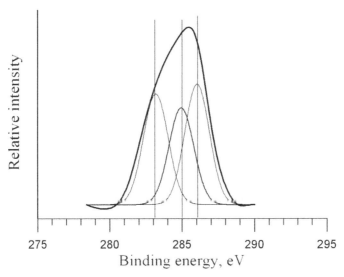

FIGURE 22.1   C1s-spectrum of the carbon nickel-containing nanostructures consisting of three components Ni-C – 283 eV, C-H – 285 eV, and C-C (sp^3) –286.2 eV.

Figure 22.2 presents the PMMA C1s-spectra. The bonds C-H (285 eV) and C-O (287 eV) characteristic of the PMMA structure prevails both in the PMMA sample containing 10^{-5}% C/Ni nanostructures and a reference sample, that is, a PMMA film. With the increase of the content of C/Ni nanostructures in PMMA starting with

10^{-4}%, the C1s-spectrum structure is changing; a structure characteristic of a C/Ni nanostructure appears: Ni–C (283 eV), C–C(sp^3) –286.2 eV and C–H (285 eV). C-H is a component characterizing the residual organic glass structure. Note that the C-O component is absent.

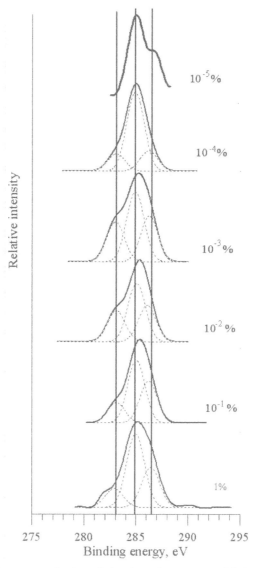

FIGURE 22.2 C1s-spectra of polymethylmethacrylate nano-modified with carbon nickel containing nanostructures in the amount of 10^{-5}%; 10^{-4}%; 10^{-3}%; 10^{-2}%; 10^{-1}%; 1%.

With the further increase of the nanostructure concentration to 10^{-3}% and 10^{-2}%, the C-H component decreases and the Ni-C and C-C (sp^3) components characteristic of the nanostructures grow. The polymer modification is observed till the content of the C/Ni nanostructures reaches 1%. With the further increase of the nanostructures concentration (3%), the process of further modification of the organic glass does not take place, which can be explained by the coagulation of the C/Ni nanoparticles. The C1s-spectrum structure becomes similar to that of the C1s-spectrum of the non-modified PMMA. Thus, we can judge about the degree of the polymer nano-modification by the ratio of the Ni-C and C–C bonds to the C–H bonds in the C1s-spectrum (see Table 22.1).

TABLE 22.1 Positions of the Maxima of the C1s-Spectrum Components for Polymethylmethacrylate the Error in the Binding Energy Measurement is ±1eV

Sample	% of nanostructures introduced into polymer	$E_{bind,}$ eV C–H	E_{bind} eV C–O	E_{bind} eV Me–C	E_{bind} eV C–C (sp^3)	$I_{C-C} + I_{Me-C} / I_{C-H}$ Degee of nano-modification
PMMA	10^{-5}	285	287	–	–	0
(organic	10^{-4}	285	–	283.0	286.2	0.5
glass)	10^{-3}	285	–	283.0	286.2	1.3
	10^{-2}	285	–	283.1	286.1	1.2
	10^{-1}	285	–	283.0	–	1,1
	1	285	–	283	286.2	0.9
	3	285	287	–	-	0

The modification reaches the maximal degree when the content of nanostructures is 10^{-3}–10^{-2}%, and with the growth of the degree of the modification the intensity of the C-C component with the sp^3-hybridization of the carbon valence electrons is increasing relative to the Ni C component intensity

The disappearance of the component characterizing the C-O bond in the C1s-spectrum indicates the possibility of the breakage of this chemical bond in the studied polymer and the formation of strong bonds between the polymer atoms and the surface atoms of the nanostructures. In this case, the nanostructures become centers of a new structure formed.

Our attention has been attracted by the fact that in the C1s-spectrum of the polymer modified with the C/Ni nanostructures the ratio of the intensities of the Ni-C and C-C components to that of the C-H component characterizes the degree of the polymer nano-structuration (see Table 22.1).

In Refs. [4–7], the dependence was studied of the degree of the modification of polymers with carbon copper-containing nanostructures (C/Cu nanostructures), which had different molecular structures and the content of oxygen bound to carbon (polycarbonate, PMMA, and polyvinyl alcohol). It was also shown that for the obtaining of the maximal degree of the modification of the polymers under study the 10^2–10^{-3}% of C/Cu nanoforms was required. In that case, the C1s-spectrum structure of the nano-modified polymers formed similar to that of the C/Cu nanoforms. In PMMA, the beginning of the modification was observed at 10^{-4}% of the nanostructures. At the content of the C/Cu of 10^{-1}%, the modification of the polymers was not observed. Thus, regardless of metal and the nanostructure form, in the C1s-spectra of nano-modified polymers such regularities are observed, and only the concentration range for carbon metal-containing nanostructures, in which the polymer modification takes place, changes. When carbon nickel-containing nanostructures are used, the PMMA modification is observed in the range of 10^{-4}%-1% and in the case of carbon copper-containing nanostructures it occurs in the range of 10^{-4}%–10^{-1}%.

In Ref. [2], the influence of the nature of a metal used for the preparation of carbon metal-containing nanostructures on their morphology was studied. In accordance to the TEM (transmission electron microscopy) studies conducted in Refs. [2, 8], when copper is used, carbon nano-film structures are formed consisting of carbon fibers. When nickel is used, carbon fiber structures are formed including nanotubes. In the case of replacement carbon copper-containing nanostructures (20–25 nm) by nickel-containing ones (11 nm), the difference in the amount of the carbon metal-containing nanostructures introduced into a polymer before they start coagulating apparently is due to the nanostructures average dimension and form, which leads to different specific surface of the nanostructures (160 m^2/g in C/Cu and 251 m^2/g in C/Ni). For the C/Ni nanostructures possessing a larger specific surface in comparison with the C/Cu nanostructures, one can observe a large activity in the interaction of the nanostructures with the surroundings and the formation of a stronger chemical bond of the atoms on the nanoform surface with the atoms of the surroundings. The increase of the content of the C/Ni nanostructures by an order in comparison with the C/Cu nanostructures for the PMMA modification can be explained by the above-mentioned data and is confirmed by the quantum-chemical calculations of the energy of the interaction of a transition-metal ion and a benzene ring [2, 5]. The energy of the interaction of a nickel ion with a benzene ring is thrice as large as that of the copper ion interaction.

22.4 CONCLUSION

1. Mechanical and ultrasonic treatment facilitates the uniform distribution of carbon metal-containing nanostructures throughout the bulk of the modified medium and the breakage of the coagulated nanostructure into separate nanostructures.

2. The maximal degree of the modification of the studied polymer is observed when the content of carbon nickel-containing nanoforms is in the range of

$10^{-3}-10^{-2}\%$. In this case the nearest surrounding of carbon atoms in the nano-modified polymer is formed similar to the structure of a nanoform.

3. A high percentage content of the nanostructures can lead to the coagulation of the nanostructures and the absence of the polymer modification. This is observed when in the studied polymer the content of carbon nickel-containing nanostructures is more than 1%.

4. When copper is replaced by nickel in carbon metal-containing nanostructures, the difference in the beginning of the coagulation of the nanostructures apparently is due to a smaller average dimension of the carbon nickel-containing nanostructures and a larger activity of their interaction with the surroundings.

KEYWORDS

- **Carbon nickel-containing nanostructures**
- **Modification**
- **Polymethylmethacrylate**
- **Satellite structure of C1s-spectra**
- **X-ray photoelectron spectroscopy**

REFERENCES

1. Trapeznikov, V. A., Shabanova, I. N., et al. (1986). New Automated X-ray Electron Magnetic Spectrometers, a Spectrometer with Technological Adapters and Manipulators, a Spectrometer for Melts Investigation, Izvestia AN SSSR, Seria fizicheskaya, *50(9)*, 1677–1682.

2. Kodolov, V. I., & Khokhryakov, N. V. (2009). Chemical Physics of the Processes of the Formation and Transformations of Nano structures and Nanosystems, IzhGSHA, Izhevsk, in Two Volumes, *1*, 360, *2*, 415p.

3. Makarova, L. G., Shabanova, I. N., & Terebova, N. S. (2005). The Use of the Method of X-ray Photoelectron Spectroscopy for Investigating the Chemical Structure of Carbon Nanostructures, Zavodskaya Laboratoria, Diagnostika Materialov, *71(5)*, 26.

4. Shabanova, I. N., Kodolov, V. I., Terebova, N. S., Ryabova, V. I., Sapozhnikov, G. V., & Obukhov, A. V. (2013). X-ray Photoelectron Study of the Influence of Minute Additions of Carbon Metal-Containing Nanostructures on the Degree of the Polycarbonate Modification, Khimicheskaya Fizika i Mezoskopia, *15(4)*, 570–575.

5. Shabanova, I. N., Kodolov, V. I., Terebova, N. S., Sapozhnikov, G. V., Poleto, Ya A V., Pershin, Yu V., & Ryabova, B. I. (2013). X-ray Photoelectron Investigation of the Influence of the Content of Carbon Metal-Containing Nanostructures and their Activity on the Polymer Modification, Proceedings of the 4th International Conference "From Nanostructures, Nanomaterials and Nanotechnologies to Nanoindustry," 110.

6. Shabanova, I. N., Terebova, N. S., & Sapozhnikov, G. V. (2013). The Study of the Mechanism of the Influence of a Minute Amount of Carbon Metal-Containing Nanoforms and their Activ-

ity on the Change of the Polymer Structure, Sbornik dokladov XXI Vserossiiskoi konferentzii "X-ray and Electron Spectra and Chemical Bond," 115.

7. Shabanova, I. N., Terebova, N. S., & Sapozhnikov, G. V. (2013). XPS Study of the Influence of Minute Additions of Carbon Metal-Containing Nano forms on the Polymer Structure, Abstract Book of 15th European Conference on Applications of Surface and Interface Analysis, 178.

8. Shabanova, I. N., Kodolov, V. I., Terebova, N. S., & Trineeva, V. V. (2012). X-ray Photoelectron Spectroscopy in the Investigation of Carbon Metal-Containing Nano systems and Nano structured Materials, Moskva-Izhevs, Iz-vo "Udmurtski Universitet," Institut Komputernikh Issledovani, 250p.

CHAPTER 23

DEPENDENCE OF THE DEGREE OF NANOMODIFICATION OF POLYMERS ON THE CONTENT OF OXYGEN ATOMS IN THEIR STRUCTURE

I. N. SHABANOVA and N. S. TEREBOVA

CONTENTS

23.1 INTRODUCTION

High structure-forming activity of nanostructures allows their use for modification of materials. It is known that the modification of different materials with minute amounts of nanostructures improves material performance characteristics. At the same time, the mechanism of such influence of nanoforms on the structure and properties of materials has not been fully clarified yet. In this chapter, the dependence of the nano-modification of polymer systems on the content of oxygen in their structure is studied, and the explanations of the process are presented based on the XPS results using the modification of polymer systems as an example.

23.2 EXPERIMENT PART

Carbon metal-containing nanostructures are multi-layered nanotubes formed in a nanoreactor of a polymer matrix in the presence of the 3d-metal systems. Nano-carbon structures are prepared by an original procedure in the conditions of low temperatures (not higher than 400 °C), which is pioneer investigations [1].

The method for the modification of polymer with carbon metal-containing nano-structures includes the preparation of fine-dispersed suspension (FDS) based on the solution of polymer in methylene chloride. Then, carbon metal-containing nano-structures are added into the prepared FDS. For the refinement and uniform distribution of carbon metal-containing nanostructures, the prepared mixture is treated with ultrasound. To prepare a film, the solvent is evaporated from the mixture at heating to 100 °C.

The XPS investigations were conducted on an X-ray electron magnetic spectrometer with the resolution 10^{-4}, luminosity 0.085% at the excitation by the AlKα line 1486.5 eV in vacuum 10^{-8}–10^{-10} Pa. In comparison with an electrostatic spectrometer, a magnetic spectrometer has a number of a advantages connected with the construction capabilities of X-ray electron magnetic spectrometers which are the constancy of luminosity and resolution independent of the energy of electrons, high contrast of spectra and the possibility of external actions on a sample during measurements [2].

The study of the variations of the C1s-spectrum shape lies in the basis of the investigations of the change of the polymer structure during nano-modification.

The identification of the C1s-spectra with the use of their satellite structure has been developed allowing the determination of the chemical bond of elements, the nearest surrounding of atoms and the type of sp-hybridization of the valence electrons of carbon in nano-clusters and materials modified with them [3–5].

Since the surface of nanostructures has low reactivity, for increasing the nano-structure surface activity functionalization is used, that is, the attachment of certain atoms of sp- or d-elements to the atoms on the nanostructure surface and the formation of a covalent bond between them; the functionalization results in the formation of an interlink between the atoms of the nanostructure surface and the atoms of material.

The XPS study has shown [6] that the formation of the covalent bond between the atoms of the functional sp-groups and the atoms on the nanostructure surface is influenced by the electro negativity of the atoms of the components and the closeness of their covalent radii. Thus, it is most probable that the functionalization of nanostructures leads to the formation of the bond between phosphorus atoms and nanostructure d-metal atoms (Fe, Ni, Cu) and between nitrogen or fluorine atoms and carbon atoms of the nanostructures.

23.3   RESULTS AND DISCUSSION

The changes of the structure of the organic glass, polycarbonate and polyvinyl alcohol modified with different amounts (10^{-5}, 10^{-4}, 10^{-3}, 10^{-2}, 10^{-1}%) of carbon copper-containing nanostructures have been studied.

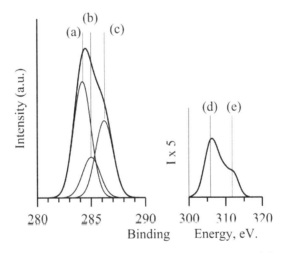

FIGURE 23.1 The XPS C1s-spectrum of carbon copper-containing nanostructures consisting of three components: (a) C–C (sp^2) – 284 eV; (b) C–H – 285 eV; (c) C–C (sp^3) – 286.2 eV, and the satellite structure; (d) satellite (sp^2); (e) satellite (sp^3).

Figure 23.1 presents the C1s-spectrum of the carbon copper-containing nanostructure, which consists of three components C–C (sp^2) – 284 eV, C–H – 285 eV, and C–C (sp^3)–286.2 eV. The presence of a small amount of the C–H component indicates an incomplete synthesis of nanostructures from the polymer matrix. The ratio of the maxima intensities of C–C (sp^2) and C–C (sp^3) depends on the dimension of the nanostructure: the larger is the surface area compared to the volume, the higher is the C1s-spectrum component with the sp³-hybridization of valence electrons [3].

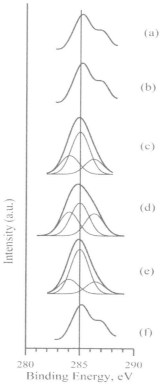

FIGURE 23.2 The XPS C1s-spectra of polymethylmethacrylate: (a) reference sample; (b) nanomodified with carbon copper-containing nanostructures in the amount of 10^{-5}%; (c) nanomodified with carbon copper-containing nanostructures in the amount of 10^{-4}%; (d) nanomodified with carbon copper-containing nanostructures in the amount of 10^{-3}%; (e) nanomodified with carbon copper-containing nanostructures in the amount of 10^{-2}%; (f) nanomodified with carbon copper-containing nanostructures in the amount of 10^{-1}%.

Figure 23.2 shows the C1s-spectra of polymethylmethacrylate (chemical formula $[CH_2C(CH_3)(CO_2CH_3)]n$). In the reference sample, which is the film of organic glass, the bonds C–H and C–O (287 eV) prevail characterizing the organic glass structure. With increasing content of carbon copper-containing nanostructures in organic glass, starting from 10^{-4}%, the C1s-spectrum structure changes and the structure characteristic of carbon copper-containing nanostructure appears, namely, C–C (sp²) and C–C (sp³). The C–H component characterizes the remnants of the organic glass structure. Note that the C–O component is absent in the C1s-spectrum.

During the organic glass modification, with increasing concentration of the nano-structure up to 10^{-3}%, in the C1s-spectrum the C–H component is decreasing and the C–C (sp²) and C–C (sp³) components characteristic of nanostructures are growing, that is, the degree of the polymer modification is growing. With a further increase in the nanostructure concentration (10^{-2}%), the degree of the polymer modification decreases and at the nanostructure content of 10^{-1}% the modification is absent. The structure of the C1s-spectrum becomes similar to that of the C1s-spectrum of the unmodified organic glass. This can be explained by the processes of the nanostruc-ture coagulation-taking place when the nanostructure content in the polymer is high. Consequently, it is possible to judge about the degree of the polymer nanomodifica-tion by the ratio of C–C–bonds to C–H–bond in the C1s-spectrum. Maximal modi-fication takes place at the nanostructure content of 10^{-3}%.

Similar results have been obtained for the nanomodification of polycarbonate (chemical formula $-[-OArOC(O)OR-]_n-$, $Ar-C_6H_4-$; $R-(CH_2)$]. The C1s-spectrum of the reference sample of polycarbonate containing a large amount of oxygen also consists of two components C–H (285eV) and C–O (287eV); however, the relative intensity of the C–O component is significantly larger than that observed for organic glass. When the content of nanostructures is in the range of 10^{-5}%–10^{-2}%, in the C1s-spectrum the structure characteristic of carbon copper-containing nanoform ap-pears. In contrast to organic glass, in polycarbonate the change of the structure starts at the nanoform content of 10^{-5}%. In polycarbonate, at the nanostructure content of 10^{-1}%, no changes are observed in the polymer structure.

Figure 23.3 shows the C1s-spectra of polyvinyl alcohol (chemical formula $[CH_2CH\,(OH)]_n$) having in its structure the least content of oxygen in comparison with the other polymers under study. Similar to the reference samples of polycar-bonate and organic glass, the C1s-spectrum of the reference sample of PVA (a PVA film) contains two components C–H and C–O; however, the C–O component is less intensive compared to C–H. The change of the PVA structure is observed only when it is modified with carbon copper-containing nanostructures, the content of which is in the range of 10^{-3}%–10^{-2}%. When nanoforms are added into the PVA solution, in the C1s-spectrum structure the components appear which are characteristic of the C1s-spectrum of carbon copper-containing nanostructure (Fig. 23.1). When in PVA the nanostructure content is 10^{-1}%, no changes are observed in the polymer structure similar to the above-mentioned polymers.

At the higher content of the nanostructures (1%) in the PVA polymer nano-modification is also absent; however in contrast to the samples with the content of nanostructures of 10^{-1}%, the absence of oxygen atoms (there is no component C–O in the C1s-spectrum) and the decrease of the component characterizing the C–H bond are observed. Consequently, the PVA carbonization takes place in the pres-ence of the nanostructures as catalysts of the reaction in which the bonds between carbon and oxygen and hydrogen are broken and deoxygenation and dehydrogen-ization and the formation of C–C bonds occur. The C1s-spectrum consists of three components C–C (283, 2 eV), C–C with sp²-hybridization (284, 2 eV) and C–H (285 eV). The presence of a small amount of the C–H component in addition to the C–C component indicates incomplete carbonization. Since the binding energy of the C–C component (283, 2 eV) is smaller than that of the C–C component with the sp²-hybridization of the valence electrons (284, 2 eV), the formation of the C–C component with sp-hybridization valence electrons of the atoms carbon can be suggested.

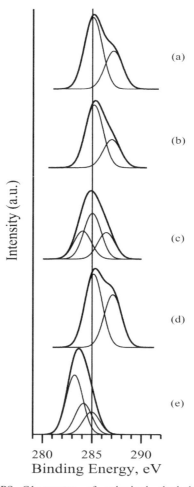

FIGURE 23.3 The XPS C1s-spectra of polyvinyl alcohol: (a) reference sample; (b) nanomodified with carbon copper-containing nanostructures in the amount of 10^{-4}%; (c) nanomodified with carbon copper-containing nanostructures in the amount of 10^{-3}%; (d) nanomodified with carbon copper-containing nanostructures in the amount of 10^{-1}%; (e) nanomodified with carbon copper-containing nanostructures in the amount of 1%;

A quite definite amount of the nanostructures is required for the modification of each of the studied polymers, that is, for changing their structure. The XPS studies show that the smaller is the number of oxygen atoms bound to carbon in the polymer, the larger is the amount of nanoparticles necessary for changing the polymer structure. For polycarbonate, it is 10^{-5}%, for organic glass–10^{-4}%, and for polyvinyl alcohol–10^{-3}%. It is associated with different reactivity of the C–O bond in polymers. When in the polymer the amount of nanoparticles is in the range from minimal up to 10^{-1}%, the C1s-spectrum changes; the component C–O (277.0 eV) disappears and the components C–C (sp^2) and C–C (sp^3) characteristic of a carbon metal-containing nanoform appear. The comparison of the structure of the studied polymers show high reactivity of the C–O bond in the CO$_3$ group in polycarbonate; the C–O bond reactivity decreases in the CO$_2$ group in organic glass, and its further decrease is observed in the CO group in polyvinyl alcohol. The disappearance of the component characterizing the C–O bond in the C1s-spectrum indicates the possibility of the replacement of this group of atoms by nanoparticles, i.e., the formation of strong bonds between the polymer atoms and the atoms of the nanostructure surface. In this case, the polymer gains the structure-forming activity of nanostructures, which are the centers of a new appearing structure. The breakage of the chemical bond of the C–O group with the nearest environment of the polymer atoms due to a different reactivity of the C–O bond takes place at different minimal contents of nanoparticles in the polymers.

The investigation of Vaseline oil (chemical formula [CH$_2$]$_m$), which does not contain oxygen atoms in its structure shows the absence of the Vaseline oil modification at any content of nanoforms. The C1s spectrum has the only C–H component similar to the reference sample.

Based on the results obtained it can be suggested that the larger is the content of oxygen atoms in the bond with carbon in the studied polymers, the larger is a change in the polymer structure and the larger is the formation of the regions in the polymer structure, which have a similar structure to that of carbon metal-containing nanoforms. The smaller is the number of oxygen atoms in the starting polymer, the larger is the amount of nanoforms necessary for the polymer structurization and the transformation of the polymer structure into the one similar to the nanoform structure during the modification.

The experimental data show that the modification of the polymers occurs in the range from room temperature up to 100°C; at higher temperatures the modification is absent because the polymer decomposition starts.

23.4   CONCLUSION

In this chapter, it is shown that the degree of the nanostructure influence on the interaction with a polymer is determined by the content of nanostructures and their activity in this medium. The temperature growth blocks the development of self-organization in the medium. Thus, for describing the medium structurization process under the nanostructure influence it is necessary to enter some critical parameters, namely, the content and activity of nanostructures and critical temperature.

The change of the polymer structure is accompanied by the change of their technological properties: the tensile strength of the studied films improves by 13%, the surface electrical resistance decreases by a factor of 3.3, and the transmission density of the films increases in the region close to an infrared one, which leads to an increase in the polymer heat capacity.

Thus, the X–ray studies show that:

1. For obtaining the maximal degree of the modification of the studied polymers it is necessary that the content of carbon copper-containing nanoforms in them would be $\sim 10^{-3}\%$. In this case, the structure of the nanomodified polymers changes and becomes similar to the structure of the nanoform.
2. The larger is the content of oxygen atoms bound to carbon atoms in the polymers; the larger is the degree of the polymer modification and the smaller is the percentage of nanoforms required for the beginning of modification.
3. High percentage of nanostructures can lead to the nanostructure coagulation and the absence of the polymer modification. In the studied polymers, it is observed at the nanostructure content of $10^{-1}\%$.
4. When oxygen atoms bound to carbon atoms are absent in the polymer structure, modification is not observed at any content of nanostructures.
5. At the higher content of nanostructures (1%), in the studied polymers the process of partial carbonization is observed, that is, the removal of O and H atoms and the formation of the C–C bonds with sp-hybridization of the valence electrons.

KEYWORDS

- **Carbon copper-containing nanostructures**
- **Functionalization**
- **Modification**
- **Polycarbonate polyvinyl alcohol (PVA)**
- **Polymethylmethacrylate (organic glass, PMMA)**
- **Satellite structure of C1s-spectra**
- **X-ray photoelectron spectroscopy (XPS)**

REFERENCES

1. Kodolov, V. I., & Khokhryakov, N. V. (2009). Chemical Physics of the Processes of the Formation and Transformations of Nanostructures and Nanosystems, Iz-vo IzhGSHA, Izhevsk, in two Volumes, Vol. 1, *360(2),* 415.
2. Shabanova, I. N., Dobysheva, L. V., Varganov, D. V., Karpov, V. G., Kovner, L. G., Klushnikov, O. I., Manakov, Yu G., Makhonin, E. A., Khaidarov, A. V., & Trapeznikov, V. A. (1986). Izv AN SSSR, Ser., Fiz., *50(9),* 1677.
3. Makarova, L. G., Shabanova, I. N., & Terebova, N. S. (2005). Zavodskaya Laboratoria, Diagnostics of Materials, *71(5),* 26.
4. Makarova, L. G., Shabanova, I. N., Kodolov, V. I., & Kuznetsov, A. P. (2004). J. Electr. Spectr. Rel. Phen, *137–140,* 239.
5. Shabanova, I. N., Kodolov, V. I., Terebova, N. S., & Trineeva, V. V. (2012). X-ray Photoelectron Spectroscopy in the Investigation of Carbon Metal-Containing Nanosystems and Nano-Structured Materials, Iz-vo "Udmurtski Universitet, " Izhevsk, 250p.
6. Shabanova, I. N., & Terebova, N. S. (2012). Journal of Nanoscience and Nanotechnology, *12(11),* 8841.

CHAPTER 24

FEATURES OF DEFORMATION RELIEF OF THE SURFACE OF THE METAL STRIP, OBTAINED BY THE HIGH PRESSURE SHEAR

A. V. SISANBAEV, A. A. DEMCHENKO, M. V. DEMCHENKO, A. V. SHALIMOVA, L. R. ZUBAIROV, and R. R. MULYUKOV

CONTENTS

24.1   INTRODUCTION

One way to obtain nanostructured metals is plastic deformation, which is based on the formation of a highly fragmented and disoriented structure due to large shear deformation [1, 2]. Torsion under high pressure (THP) is one of the effective methods of shear mega plastic deformation [3]. In the method of THP shear deformation occurs due to torsion sample movable peen. Due to the specific method of THP (see Fig. 24.1) has a radial inhomogeneity of the degree of deformation in different areas. At the same time, one cycle of deformation can be obtained by different structures of the radial dependence of the shear deformation, but deformation general parameters: pressure, temperature, and the angular velocity of shear. Attached coaxially with the center of pressure of the sample, usually reaching several GPa, plays an important role. First, it creates in the central part of the sample area quasi-hydrostatic compression, preventing the destruction of a homogeneous sample. Second, it increases the frictional force between the peen and the sample.

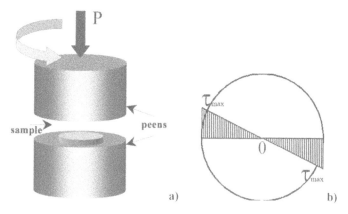

FIGURE 24.1   The method of THP: (a) scheme and (b) the distribution of shear stresses in the sample.

Due to the large friction force, torque is transmitted from the mobile peen sample. In cross-sections of the sample shear stresses occur, leading to pure shear. Since the pair peen-sample work closely with each other work surfaces, there is a need to provide deformation relief features associated with the deformation behavior is investigated metal.

The aim of this study was to estimate the quantitative characteristics of the deformation relief "cramped" surface of the sample and mobile peen after THP at different structural levels.

24.2 MATERIALS AND METHODS OF EXPERIMENTS

The object of investigation was selected metal alloy with shape memory Ti-49, 8% Ni. Sample diameter d = 10 mm and a thickness δ = 0.2 mm offset deformed at a temperature T = 300K and P = hydrostatic pressure of 5 GPa. Pure shear deformation is determined by the formula $\gamma - \pi N\delta / \delta$, where N – number of revolutions of the Stithy. One revolution of the peen with an angular velocity ω = 1 rev/min allowed to reach the periphery of the sample degree logarithmic deformation E_{max} = 5. For the original homogeneous surface relief of the peens and their working sample surface is exposed step polishing sandpaper with different grit abrasive in descending order (for foreign classification from P180 to P2500).

For the filming of the initial deformation and surface topography of the peen and the sample used confocal laser 3D-microscope "LSM-Exciter" (Carl Zeiss, Germany) with the program 3D-analysis "ZEN" [4–11]. The scanning system allows you to create 3D–Image using up to $\sim 10^3$ shots optical sections of the surface relief height increments up to ~ 10 nm at a resolution in the plane to ~ 100 nm. At different structural levels measured parameters "dispersion" of the surface relief: local R_a (arithmetic mean deviation from the mean value of the profile) and integral RS_a (average R_a for multiple profilograms scanned area). Under different structural levels implied scale measured "variance" with a set of lenses increases "m": 5^x, 10^x, 20^x, 50^x and 100^x. That is, "meso" ~ 10 mm, "micro" ~ 1 mm and "nano" ~ 100 nm. For a more detailed analysis of the features of the deformation of the sample surface topography using an optical 3D-microscope "Axio–Imager-Vario" with the possibility of smooth changes magnification and scanning electron microscope "Merlin" (Carl Zeiss, Germany).

24.3 RESULTS AND DISCUSSION

The process of deformation of the samples under the THP accompanied by a distinct crackle associated with acoustic emission characteristic of the formation of micro cracks. Figure 24.2 shows laser 3D scanning the working "cramped" surfaces of the sample after the THP: a) general view of deformation relief and b) typical radial profilogram relief from center to edge.

At the micro level (Fig. 24.3) are seen as irregular deformation bands perpendicular to the shear direction, and the area with microcracks. Drawn through these areas transverse profilogram clearly registers deep depressions in places microcracks. The relief has a nonuniform intervals associated with an irregular arrangement of parallel strips deformation. The direction of the microcracks is correlated with an intermediate angle between the direction of shear deformation and stripes, which are in turn orthogonal. In our deformation conditions it is close to the angle of $\sim 45°$. Optical and electronic 3D-scan (Figs. 24.4 and 24.5) the working "cramped" surfaces of the sample after the THP at different structural levels revealed in another contrast morphology of the fine structure of accommodative transverse deformation bands.

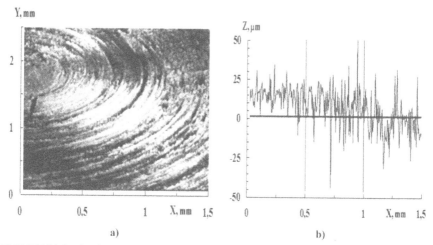

a)   b)

FIGURE 24.2   3D-laser scanning of the sample surface after the HPC: (a) general view and (b) profilogram relief from center to edge.

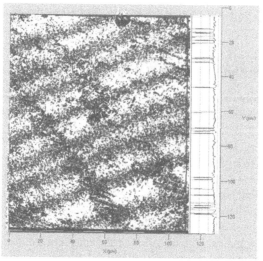

FIGURE 24.3   Laser 3D-scanning area between the center and the edge of the sample surface after the THP: irregular deformation bands perpendicular to the shear direction and transverse profilogram with deep depressions in places micro cracks.

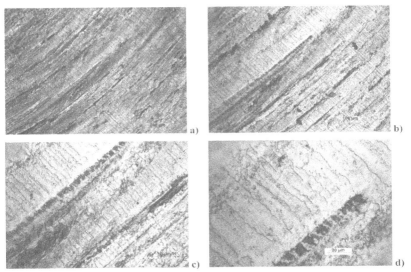

FIGURE 24.4 Optical 3D-scan region between the center and edge of the sample surface after THP different objectives: (a) 10ˣ, (b) 20ˣ, (c) 50ˣ and (d) 100ˣ.

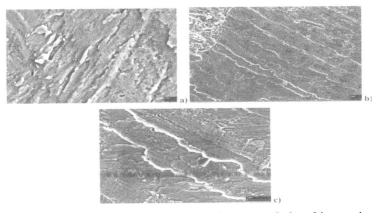

FIGURE 24.5 Electronic 3D-scan region between the center and edge of the sample surface after THP: (a) the microstructure of the border area between the circular shear zones, (b) and (c) the microstructure of the transverse deformation bands at different magnifications.

In the border area between the circular shear zones are clearly visible signs of the flow of a crystalline solid and the occurrence of deformation zones inconsistent accommodative ruptures and discontinuities. Measurements showed (see Table 24.1) that the deformation at the micro topography of the sample after the THP corresponds relief movable anvil.

TABLE 24.1 Options "Cramped" Relief Work Surfaces

Structural levels (m)	Peen RS$_a$ (R$_a$), mm		Sample RS$_a$, mm
	Initial	After THP	After THP
meso (5^x)	9.4	8.7 (7.0)	15.0
micro (10^x)	2.9	1.3 (1.0)	1.2
micro (20^x)	1.6	1.2 (1.4)	1.1
micro (50^x)	0.4	0.2 (1.2)	0.2
nano (100^x)	0.2	0.1 (0.1)	0.2

That is, in a pair of contact surfaces "peen-sample" of their dispersion relief virtually identical (within heterogeneity). Heterogeneity is clearly visible when comparing the integral and local relief dispersions. Meso – and nanoscale RS$_a$ ratio for the sample and the peen differ markedly (~2 once). Anomaly in the increased value of the sample variance relief from the meso level can be attributed to additional relief from the transverse deformation bands. Comparative analysis of the dispersions of the peen working surface topography before and after THP showed a decrease RS$_a$, that is smoothes the original topography. Figure 24.6 shows the dependence of the integrated variance of the structural level studies of surface topography pin (before and after THP) and the sample. Trend lines of these curves are the closest to the type of power dependence RS$_a$ = 65 m$^{-1.4}$ with a range of values of ~0.1–15 mm.

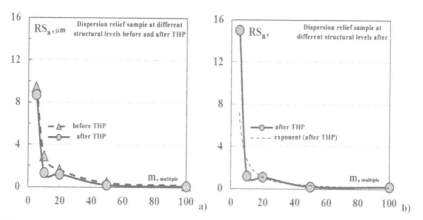

FIGURE 24.6 Dependence of the variance of the structural level studies of surface topography: (a) the peen and (b) of the sample.

24.5   CONCLUSIONS

A quantitative validation on different structural levels parameters of surface topography and the peen sample using laser scanning 3D-microscopy. Based on a comparative analysis of integrated variance relief after HPC for the peen and the sample found that meso-and nanoscale RS_a different ratio of the ~2 once. Study sample using laser scanning, optical and electron microscopy revealed the structure of the surface deformation accommodative presence of transverse deformation bands orthogonal directions shift.

KEYWORDS

- **Band deformation**
- **Deformation relief on the surface**
- **Laser scanning 3D-microscopy**
- **Mega plastic deformation**
- **Metal with a shape memory effect**
- **Microcracks**
- **Shear under high pressure**
- **Structural levels**

REFERENCES

1. Segal, V. M., Reznikov, V. I., Drobyshevskiy, F. E., & Kopylov, V. I. (1981). Plasticity Metals Working of simple Shear, Izvestiya AS USSR Metals, *1*, 115–123.
2. Akhmadeev, N. A., Valiev, R. Z., Kopylov, V. I., & Mulyukov, R. R. (1992). Formation Submicron Grained Structure in Copper and Nickel Using Intensive Shearing, Izvestiya RAS, Metals, *5*, 96–101.
3. Glezer, A. M., Sundeev, R. V., & Shalimova, A. V. (2011). The Cyclical Nature of the Type of Crystal Phase TransformationsÛAmorphous State at Mega Plastic Deformation Alloy Ti-50Ni25Cu25, Reports of the Academy of Sciences, *440(1)*, 39–41.
4. Yamshikova, S. A., Kravtsov, V. V., & Sisanbaev, A. V. (2009). Fire Protection of Metal Structures Modified Intumescent Coatings, Quality Management in the Oil and Gas Complex, *2*, 41–43.
5. Kruglov, A. A., Rudenko, O. A., & Sisanbaev, A. V. (2009). Super Plastic Forming Nano Structured Titanium Alloy Sheets, Proc. Intern Works, STC "Progressive Methods and Technological Equipment of Metal Forming Processes" Street Petersburg.BSTU "Voenmeh" of Name Ustinova, D. F., 92–95.
6. Yamshikova, S. A., Kravtsov, V. V., Sisanbaev, A. V., & Iskanderov, A. R. (2009). Increased Fire Resistance of Metal Structures by Applying Intumescent Compositions, Proc. International Works, STC "Actual Problems of Engineering, Natural Sciences and Humanities" Ufa Publisher UGNTU Edition 4, 112–114.

7. Bakiyev, A. V., Sandakov, V. A., & Sisanbaev, A. V. (2010). Structural Nature of the Degradation of the Mechanical Properties of the Metal Gas Pipeline Supply System, Rostehnadzor, Ural Region, *2*, 12–15.

8. Kruglov, A. A., Rudenko, O. A., & Sisanbaev, A. V. (2011). The Temperature Dependence of the Deformation Relief Nanostructured Titanium Alloy After Super Plastic Forming Advanced Materials, *12*, Special Edition, 258–261.

9. Demchenko, A. A., Demchenko, M. V., Sisanbaev, A. V., Naumkin, E. A., & Kuzeev, I. R. (2012). Relationship Deformation Relief Surface and the degree of Damage of Steel at Low-Cycle Loading, Chemical Physics and Mesoscopy, *14(3)*, 426–429.

10. Demchenko, A. A., Demchenko, M. V., Sisanbaev, A. V., & Kuzeev, I. R. (2012). Studies of the Fractal Dimension of the Deformation of Steel Surface by Laser Scanning, Chemical Physics and Mesoscopy, *14(4)*, 569–573.

11. Demchenko, A. A., Demchenko, M. V., Sisanbaev, A. V., Naumkin, E. A., & Kuzeev, I. R. (2013). Study the Relationship of Deformation Relief and the Degree of Damage of Steel, Factory Laboratory, Diagnosis Materials, *79(2)*, 42–44.

CHAPTER 25

THE STRUCTURAL ANALYSIS OF THE MICROHARDNESS OF PLASTICIZED POLYPROPYLENE AND ITS COMPOSITES

A. L. SLONOV, G. V. KOZLOV, A. K. MIKITAEV, and G. E. ZAIKOV

CONTENTS

25.1   INTRODUCTION

At present we know [1–3] that the micro hardness *HB* is a property that is sensitive to the morphological and structural changes in polymeric materials. For composite materials an additional potent factor is the presence of the filler, whose micro hardness is much higher than in the polymer matrix [4]. At the introduction into polymer pointed indenters in the form of a cone or a pyramid stress state is localized in a small micro volume and it is assumed that such trials "groped" real structure of polymeric materials [5]. Due to the fact that the structure of polymeric composites is rather complicated [6], the question arises how the component of the structure responds to the indentation and for how much this reaction is modified with the introduction of filler.

Another aspect of the problem is the relationship of micro hardness determined by the results of the tests in a very localized microscopic volume, with such macroscopic properties of polymeric materials as elastic modulus E and yield strength σ_Y. At the moment there is quite a number of theoretically and empirically derived relationships between HB, E and σ_Y [7, 8].

The purpose of this paper is to describe the micro hardness within the above stated models and finding its relationship with mechanical and structural characteristics on the example of plasticized polypropylene and composites based on it, filled with calcium carbonate (chalk), and talc.

25.2   RESULTS AND DISCUSSION

Polypropylene homopolymer "Stavrolen" grade PPG 1035–08 (PPS), plasticized ethylene-vinyl acetate copolymer (EVA) grade 12206–007, containing up to 20% vinyl acetate and having a melt flow index of 1 g/10 min at a temperature of 463 K was used. The content EVA was varied in the range of 0–30 wt. %.

Besides, two series of composites based on plasticized polypropylene grades were used. For the first series of composites as the matrix polymer was used PPS containing EVA W_c =0, 10, 15 and 20 wt. %. As a filler calcium carbonate ($CaCO_3$) hydrophobized with stearic acid grade M90T, produced by "Ruslime," with an average size of particles 1 micron was used.

For the second series of composites as a matrix polymer mixture was used PPS and a block copolymer of propylene and ethylene (ethylene content of about 5%), grade PP 8300N, production Ltd. "Nizhnekamskneftekhim" (PPN). The ratio of the PPS/PPN was 60:40 and 70:30 by weight. As a filler A7C talc grade, with an average particle size a 2 microns, and talc having an average particle size of 4, 9 microns, Turkey production (TT) was used.

All test materials are prepared by mixing the polymer components on a twin screw melt extruder model PSHJ-20 production Jangsu Xinda Science Technology (China). Mixing was performed at temperature of 463–503 K and screw speed of

150 rev/min. Test samples were obtained by method of injection molding for injection molding machine Polytest Company Ray-Ran (Great Britain) at a temperature of 503 K and pressure of 43 MPa.

Mechanical uniaxial tensile tests were performed on the samples in the form of double-sided blade with the dimensions according to GOST 112–62–80. Testing was carried out on a universal testing machine Gotech Testing Machine CT-TCS 2000 production Taiwan, at temperature of 293K.

Measuring of micro hardness *HB* Shore (scale D) was made according to GOST 24–621–91 on a hardness tester model "Hildebrand," German production. Measurements of micro hardness were carried out after 1 second (maximum value *HB*) and 15 seconds (*HB* value in after stress relaxation) stay of the sample under load. At least five micro hardness measurements at different locations of the sample surface at a distance of no less than 6 mm from the previous measurement were achieved. The result should be the arithmetic mean of at least five measurements. Samples for measurement of *HB* had a cylindrical shape with a diameter of 40 mm and a height of 5 mm.

Let us consider the interconnection of micro hardness *HB* and other mechanical characteristics, particularly yield stress σ_Y for polymeric materials studied. Tabor [9] found for metals, which are considered as rigid perfectly plastic solids, the following relation between HB and σ_Y:

$$\frac{HB}{\sigma_Y} \approx c \, , \qquad (1)$$

where *c* is constant, which is approximately equal to 3.

The Eq. (1) means that in the attached tests on micro hardness the pressure under indenter is higher than the yield stress in the quasistatic tests because of the tightness imposed on undeformed polymer surrounding the indenter. However, the authors [3, 7, 8, 10, 11] showed that the value *c* may differ significantly from the 3 and vary within a wide range: 1, 5–30. In Ref. [11], it was found out that for high-density polyethylene composites/calcium carbonate, depending on the strain rate $\dot{\varepsilon}$ and the type of quasi-static tests, which determine the value of σ_Y (tension or compression), the value varies within 1.80–5.83. K_c=3 ratio HB/σ_Y approaches only at the minimum $\dot{\varepsilon}$ and at using σ_Y values obtained in the compression test. Therefore, in the Ref. [11], it was concluded that the quantity *c*=3 can be obtained only at speeds comparable to the test strain and on micro hardness and quasistatic tests with no destruction of interfaces polymer filler.

For distribution of analysis on a wider range of solids it was proposed to consider the role of elasticity in the process of indentation. For a solid body with an elastic modulus *E* and Poisson's ratio n Hill the following equation was obtained [7]:

$$\frac{HB}{\sigma_Y} = \frac{2}{3}\left[1 + \ln\frac{E}{3(1-v)\sigma_Y}\right], \qquad (2)$$

and empirical equation of Marsh has the form [7]:

$$\frac{HB}{\sigma_Y} = \left(0,07 + 0,6\ln\frac{E}{\sigma_Y}\right), \qquad (3)$$

The Eqs. (2) and (3) allow to estimates HB/σ_Y for the considered polymeric materials provided known E and σ_Y, and the value n can be calculated by using the ratio [5]:

$$\frac{\sigma_Y}{E} = \frac{1-2v}{6(1+v)}, \qquad (4)$$

Let us consider the physical nature of the deviation ratio HB/σ_Y of constant $c \approx 3$ in the Eq. (1), using for this purpose the Eqs. (2) and (3). As it is known [12], fractal (Hausdorff) dimensions d_f structure of polymeric materials can be calculated according to the equation:

$$d_f = (d-1)(1+v), \qquad (5)$$

where d is the dimension of the Euclidean space, which is considered a fractal (obviously, in this case $d = 3$).

The combination of Eqs. (2)–(5) leads to the following formula [13]:

$$\frac{HB}{\sigma_Y} = \frac{2}{3}\left\{1 + \ln\left[\frac{2d_f}{(4-d_f)(3-d_f)}\right]\right\} \qquad (6)$$

and

$$\frac{HB}{\sigma_Y} = \left[0,07 + 0,6\ln\left(\frac{3d_f}{3-d_f}\right)\right], \qquad (7)$$

for the case of three-dimensional Euclidean space.

The experimental HB/σ_Y after a stay of 15 sec specimen (sample) under load, and calculated according to Eqs. (6) and (7) $(HB/\sigma_Y)^T$ ratio values of the micro hardness and tensile stress, as well as the structural and mechanical characteristics of the polymeric materials are shown in Table 25.1. As it follows from the data of Table 25.1, the results for plasticized PPS are described well by Marsh empirical equation (Eq. (7), the average discrepancy between theory and experiment D=3.5%), than the more strictly derived Hill equation (Eq. (6), D=17.6%). Let us note that a similar trend is observed for all classes of polymeric nanocomposites: particulate-filled [13] filled with organoclay [14] and carbon nanotubes [15]. However, for microcomposites, that is, composites filled with micron-sized, the opposite pattern is observed: Hill equation (Eq. (6)) gives better agreement with experiment (D=5.5%), than Marsh formula (Eq. (7), D=10.8%). Consequently, micro composites by their properties are closer to the classical elastic-plastic solids than both polymers and polymer nanocomposites.

This postulate is demonstrated visually by the data in Fig. 25.1, which shows the dependence of the ratio HB/σ_Y on the fractal dimension d_f for the considered structure of polymeric materials. As you can see, if the data for plasticized PPS are in good agreement with the theoretical curve calculated according to Eq. (7), then the data for composites PPS/CaCO$_3$, PPS-PPN/TT, PPS-PPN/A7C are shifted toward the theoretical curve calculated according to the Eq. (6). This leads to the fact that the criterion of Tabor (Eq. (1)) is fulfilled for micro composites earlier (when $d_f \approx 2.845$), than for polymers and nanocomposites (with $d_f \approx 2.95$). The last value of d_f is the maximum attainable dimension for real solids [12].

TABLE 25.1 Structural and Mechanical Properties of Plasticized PPS Composites PPS/CaCO$_3$ and PPS-PPN/Talc

Material	W_n, %	W_c, %	E, MPa	s_T, MPa	d_f	HB/σ_Y	$(HB/\sigma_Y)^T$, Eq. (6)	D, %	$(HB/\sigma_Y)^T$, Eq. (7)	D, %	$(HB/\sigma_Y)^T$, Eq. (8)	D, %
PPS	-	0	1100	29.5	2.776	2.20	2.67	21.4	2.24	1.8	2.17	1.4
	-	5	1100	27.2	2.792	2.24	2.74	22.3	2.29	2.2	2.22	0.9
	-	10	1035	26.0	2.790	2.28	2.73	19.7	2.28	-	2.21	3.1
	-	15	900	24.0	2.776	2.33	2.67	14.6	2.24	4.0	2.17	6.9
	-	20	810	21.5	2.778	2.33	2.68	15.0	2.49	6.9	2.18	6.4
	-	30	765	20.0	2.780	2.40	2.70	12.5	2.25	6.3	2.18	9.2
PPS/ CaCO$_3$	20	-	1440	25.2	2.850	2.70	3.0	11.0	2.50	7.4	2.79	3.3
	20	10	1150	22.2	2.834	2.75	2.92	6.2	2.43	11.6	2.72	1.1
	20	15	1000	21.0	2.822	2.86	2.86	-	2.39	16.4	2.66	7.0
	20	20	900	21.5	2.799	2.70	2.76	2.2	2.31	14.4	2.58	4.4
PPS-PPN /TT	20	5	1400	25.0	2.822	2.72	2.86	4.9	2.39	12.1	2.66	2.2

PPS-PPN /A7C	10	10	1200	24.0	2.830	2.71	2.90	7.0	2.42	10.7	2.70	0.4
	20	10	1400	23.6	2.854	2.75	3.02	8.9	2.51	8.7	2.81	2.2
	20	10	1435	24.7	2.852	2.63	3.0	14.1	2.50	4.9	2.80	6.5

Note: *HB* value was measured after 15 stay under load.

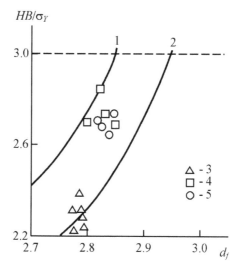

FIGURE 25.1 Dependence of the ratio HB/σ_Y after 15 stay under the load of fractal dimension structure d_f. 1, 2 – calculation according to Eqs. (6) and (7), respectively; 3–5 – experimental data for plasticized PPS (3) composites PPS/CaCO$_3$ (4) and composites based on PPS-PPN (filler – talc) (5).

$$\frac{HB}{\sigma_Y} = 0,60\ln\left(\frac{3d_f}{3-d_f}\right) \tag{8}$$

The authors [15] have shown that the modified March equation in the above form (see Eq. (8)) gives an even better agreement with the experiment than the original Eq. (3). In Table 25.1, the comparison of experimental and calculated according to Eq. (8) units HB/σ_Y is adduced. As one can see, for plasticized PPS a good agreement (D=4.7%) is obtained, but for composites this agreement is significantly worse (D=13.6%). Since the Eq. (8) is obtained on the basis of the empirical formula (the Eq. (3)), it can be corrected by replacing the constant coefficient 0.60 for polymers and nanocomposites with the coefficient 0.69 for microcomposites. In this case, the correspondence between theory and experiment is noticeably improved (D=3.4%, Table 25.1).

Finally let us consider dependences of HB/σ_Y(d_f), obtained after 1 sec stay sample load (maximum values *HB*) for the considered of polymeric materials, which are shown in Fig. 25.2. In this case, a quick increase in the ratio HB/σ_Y with the increasing plasticizer content W_c (indicated by the arrow in Fig. 25.2) for plasticized PPS, showing the transition from the viscoelastic behavior to elastoplastic, is observed whereas for composites, this dependence corresponds to the Eq. (6). The comparison of the data in Figs. 25.1 and 25.2 shows that the intensity flowing relaxation in case of plasticized PPS quickly reduces the value HB/σ_Y up to the values typical for a viscoelastic solid, whereas for composites based on plasticized PPS this decline is pronounced much more slightly.

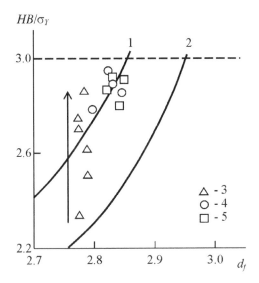

FIGURE 25.2   Dependence of the ratio HB/σ_Y after 1 to stay under the load of the fractal dimension of the structure d_f designations are the same as in Fig. 25.1.

25.3   CONCLUSIONS

Thus, the results of this work have shown that the ratio of micro-hardness and yield stress is determined only by the structural state of polymer material, which is characterized by its fractal dimension. Marsh empirical relation (eq. (3)) gives the best agreement with experiment for polymers and nanocomposites (viscoelastic solids), and more strictly Hill obtained formula (Eq. (2)) for micro composites (elastoplastic solids). Criterion of Tabor for elastoplastic solids is realized at lower fractal dimensions of the structure than for viscoelastic ones.

KEYWORDS

- **Composite**
- **Fractal dimension**
- **Micro-hardness**
- **Polypropylene**
- **Structure**

REFERENCES

1. Balta-Calleja, F. J., & Kilian, H. G. (1988). New Aspects of Yielding in Semi Crystalline Polymers Related to Micro-structure, Branched Polyethylene, Colloid Polymer Sci., *266(1)*, 26–34.
2. Balta-Calleja, F. J., Santa Cruz, C., Bayer, R. K., & Kilian, H. G. (1990). Micro hardness and Surface Free Energy in Linear Polyethylene, the Role of Entanglements, Colloid Polymer Sci., *268(5)*, 440–446.
3. Aloev, V. Z., & Kozlov, G. V. (2002). Fizika Orientazionnykh Yavleniy v Polimernyikh Materialach, Nalchik Poligrafservis i T, 288 s.
4. Perry, A. J., & Rowcliffe, D. J. (1973). The Micro hardness of Composite Materials, J. Mater Sci., Lett., *8(6)*, 904–907.
5. Kozlov, G. V., & Sanditov, D. S. (1994). Angar Monicheskie Effekty i Fiziko-Mehanicheskie Svoystva Polimerov, Novosibirsk, Nauka, 261 s.
6. Kozlov, G. V., Yanovskiy, Yu G., & Karnet, Yu N. (2008). Struktura i Svoystva Dispersno-Napolnennyih Polimernykh Kompozitov, Fraktal'nyi Analiz, M. Alyanstransatom, 363 s.
7. Kohlstedt, D. L. (1973). The Temperature Dependence of Micro Hardness of the Transition Metal Carbides, J. Mater Sci., *8(6)*, 777–786.
8. Kozlov, G. V., Beloshenko, V. A., Aloev, V. Z., & Varyuhin, V. N. (2000). Mikrotverdost' Sverkhvyso-Komolekulyarnogo Polietilena i Komponora Na Ego Osnove, Poluchennykh Metodom Tverdofaznoi Ekstruzii, Fiziko-Himicheskaya Mehanika Materialov, T, *36(3)*, 98–102.
9. Tabor, D. (1951). The Hardness of Metals, New York Oxford University press, 329p.
10. Afashagova, Z. H., Kozlov, G. V., Burya, A. I., & Zaikov, G. E. (2007). Teoreticheskaya Otsenka Mikro-Tverdosti Dispersno-Napolnennykh Polimernykh Nanokompozitov, Teoreticheskie Osnovy Khimicheskoi Tehnologii T, *41(6)*, 699–704.
11. Suwanprateeb, J. (2000). Calcium Carbonate Filled Polyethylene, Correlation of Hardness and Yield Stress, Composites, Part, AV, *31(3)*, 353–359.
12. Balankin, A. S. (1991). Sinergetika Deformiruemogo Tela M, Izd-vo Ministerstva Oborony SSSR, 404 s.
13. Kozlov, G. V., & Zaikov, G. E. (2012). Struktura i Svoystva Dispersno-Napolnennykh Polimer nykh Nanokompozitov, Saarbrücken, Lambert Academic Publishing, 112c.
14. Dzhangurazov, Zh B., Kozlov, G. V., & Mikitaev, A. K. (2013). Struktura i Svoystva Nanokompozitov Polimer/Organoglina, M Izd-vo RHTU im, Mendeleeva, D. I., 316 s.
15. Zhirikova, Z. M., Kozlov, G. V., & Aloev, V. Z. (2012). Strukturnyi Analiz Mikrotverdosti Poli-mernykh Nanokompozitov, Napolnennykh Uglero Dnymi Nanotrubkami i Nanovoloknami, Fundamentalnye Problemy Sovremennogo Materialove Deniya T, *9(1)*, 82–85.

CHAPTER 26

COMPOSITION MODIFICATION OF THE TI-6AL-4 V ALLOY SURFACE LAYERS AFTER ION-BEAM MIXING OF AL AND HEAT TREATMENT

V. L. VOROBIEV, P. V. BYKOV, S. G. BYSTROV, A. A. KOLOTOV, YA. V. BAYANKIN, V. F. KOBZIEV, and T. M. MAKHNEVA

CONTENTS

26.1   INTRODUCTION

The operating characteristics of metals and alloys comprising corrosion and erosion resistance, fatigue failure, friction, wear and crack resistance under corrosion
fatigue and a number of other properties are determined by the structure-phase condition of the surface layers [1–3]. The technique of ion implantation is a promising
method of modification of the operation properties for many construction materials
and, in particular, titanium alloys [4–6]. The given method allows us to reduce the
time and temperature of the effect on the material by tens of times, to carry out selective treatment of certain parts of the components as well as automates the process
of treatment to a certain extent. However, ion synthesis is a complicated physical-
mechanical process, in which the formation of secondary phases, their morphology,
structure and surface layers properties are determined by a complex of physical
conditions, depending on which they can change over a wide range. The implantation of the alloying elements ions into the titanium alloys surface layers by ion-beam
techniques results in the formation of inter metallide secondary phases providing
not only high mechanical properties but also good physical-chemical characteristics
(antifriction and anticorrosion).

In this connection it was of interest to study the effect of the ion-beam mixing of Al followed by post-irradiation annealing on the formation of the chemical
composition and structure-phase condition of the surface layers of the Ti-6Al-4 V
titanium alloy.

26.2   EXPERIMENTAL RESEARCH

The samples of the Ti-6Al-4 V titanium alloy were plates with the dimensions of
$9 \times 9 \times 2$ mm^3, cut out by means of spark cutting from a sheet in the as-delivered state.
The samples were mechanically polished using polishing pastes and then they were
cleaned in organic solvents. Thermal pretreatment was exposure to the temperature
of 800 °C for 1 h in the vacuum of ~10^{-4} Pa to recover the structure of the samples
and enable transition into the equilibrium state. The samples were cooled in vacuum
at room temperature.

Al deposition was carried out by magnetron sputtering on the "Cathode-1 M"
device up to the thickness of ~30 nm at the temperature of 200 °C for better adhesion
of the film and the sample. The Al film mixing was performed by the method of Ar
ion implantation in a periodic-pulse mode with the power of 30 keV, the irradiation
dose being 10^{17} ion/sm^2, the pulse current 10 mA, the frequency of succession of
length of pulses being 100 H and 1 ms, respectively, using the original ion-beam device PION-1 M on the basis of the ultrahigh vacuum chamber USU-4. The temperature of the samples during the ion irradiation was maintained with the thermocouple
and did not exceed 100 °C. After the ion-beam mixing the samples were subjected

to post-irradiation annealing at the temperature of 900 °C for 30 min in the vacuum ~10^{-4} Pa followed by cooling together with the furnace.

The examination of the topography of the samples surface was carried out by the AFM with the probe microscope SOLVER 47 PRO in the contact operation. The mean arithmetic roughness (R_a) of the surface of the samples studied was counted according to the AFM images of 15 parts of the surface with the basic size of 1×1 mkm² for every sample using of the software for processing the probe microscope data.

The chemical composition of the surface layers was investigated by X-ray photoelectron microscopy (XPEM) with the SPECS spectrometer, using the MgK_a-radiation (1253, 6 eV) together with the layer-by-layer etching of the surface by Ar ions (the calculated rate of etching ~1 nm/min).

The phase composition was determined qualitatively with the X-ray diffractometer D2 PHASER with the Bragg-Brentano geometry using a linear meter LYNX-EXE. The sample was irradiated with CuK_a-radiation, the diffraction pattern was analyzed with the help of the software module "DIFFRAC.EVA," the phases were identified applying the data base PDF-2/Release 2010 RDB of the international center for diffraction data ICDD (The International Centre for Diffraction Data).

The micro hardness of the surface layers of the samples prior and after irradiation was measured by the method of indentation of a diamond indenter using the PMT-3M device under the loading of 20 g and the sample exposure to the load for 5 sec. To increase the reliability of the result, the procedure was repeated not less than 20 times.

26.3 RESULTS AND DISCUSSION

The AFM technique examination has shown that ion-beam mixing of the film by ion implantation of Ar in the periodic-pulse mode does not result in the change of the sample surface morphology (Table 26.1). The value of the surface roughness parameter R_a for the sample in the initial state does not differ from the R_a values after Al deposition and Ar ion implantation.

TABLE 26.1 The Change of the Sample Surface Roughness of the Ti-6Al-4 V Alloy in the Initial State, after Al Deposition and After Ion-Beam Mixing

Sample	Roughness R_a, nm	CKO, nm
Ti-6Al-4 V initial state	8.4	2.9 (35%)
Ti-6Al-4 V+Al (30 nm)	8.3	2.1 (25%)
Ti-6Al-4 V+Al+Ar$^+$	7.8	2.4 (31%)

X-ray photoelectron studies of the element composition of thin surface layers in the nanometer range of the Ti-6Al-4 V alloy before and after mixing have revealed that Ar ion implantation in the periodic-pulse mode with the chosen parameters does not result in significant mixing of the Al film of ~30 nm thickness with the substrate (Fig. 26.1). It can be attributed to low energy of bombarding by Ar ions – 30 keV.

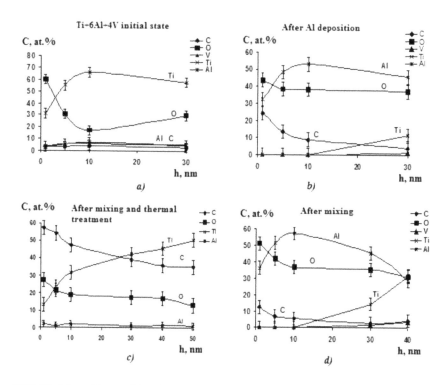

FIGURE 26.1   The profiles of elements distribution in the surface layers of the Ti-6Al-4 V titanium alloy: (a) in the initial state, (b) after Al deposition, (c) after mixing and heat treatment, and (d) after mixing.

Post-irradiation annealing of the samples after ion-beam mixing at T–900 °C for 30 min in vacuum ~10^{-4} Pa followed by cooling together with the furnace leads to the Al content decrease in thin surface layers down to trace amounts of ~1–2 at.% (Fig. 26.1c). However, the C content increased significantly, which is vividly shown in Fig. 26.2. If in the initial state and after the ion-beam mixing the C content changes within 3–13 at.%, then after the post-irradiation annealing the integral C content in the thin surface layers increased up to ~58 at.% (Fig. 26.2).

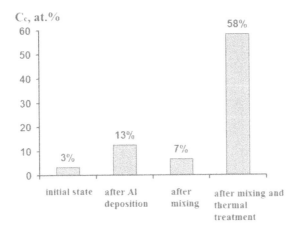

FIGURE 26.2   Integrals C content in the surface layer 30 nm in the samples of the Ti-6Al-4 V titanium alloy in the initial state, after mixing and thermal treatment.

The analysis and comparison of the X-ray photoelectron spectra of the $Ti2p_{3/2}$ lines and those of the C1 s samples in the initial state, after mixing and post-irradiation annealing revealed the formation of the carbide-like Ti compounds of the TiC, Ti_2C type (Figs. 26.3 and 26.4) in the surface layers in the latter case. It is proved by the shift of the maximum position of the $Ti2p_{3/2}$ line by 0, 8 eV with respect to its position for the initial sample (Fig. 26.3). In the initial state the binding energy of titanium is 454,0 eV, but after the post-irradiation annealing the $Ti2p_{3/2}$ line shifts towards higher values of the binding energy up to the value of 454,8 eV (Fig. 26.3). In the C1 s carbon spectra of this sample there reveals a peak with the binding energy of 281, 8 eV which does not disappear in the process of etching up to the depth of ~50 nm (Fig. 26.4). According to the reference data of the X-ray electron spectroscopy this binding energy corresponds to the C binding energy in carbides [7], in particular, the compounds of the TiC, Ti_2C types can be formed (Fig. 26.4).

The qualitative X-ray phase analysis of the diffraction patterns of the samples at different stages of treatment has shown that after the post-irradiation annealing of the sample at T=900 °C in vacuum followed by cooling together with the furnace down to the room temperature, such Ti carbides as Ti_2C, TiC and carbonitride Ti_2CN (different part 3 in Fig. 26.5) are formed in the surface layers. Meanwhile the peaks accounting for these phases are not identified (different parts 1 and 2, Fig. 26.5, respectively) in the diffraction patterns corresponding to the samples in the initial state and after ion-beam mixing of Al, qualitatively not being different from each other.

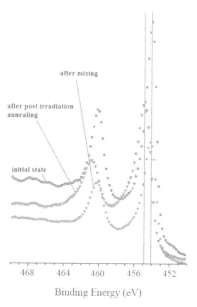

FIGURE 26.3   The Ti2p lines of the Ti-6Al-4 V titanium alloy in the initial state at the depth of ~40 nm, after mixing and post-irradiation annealing at the depth of ~40 nm.

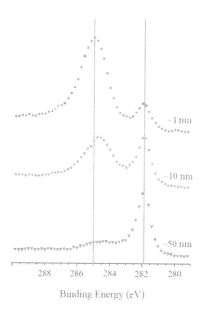

FIGURE 26.4   The C1 s carbon spectra in the surface layers of the nanometer scale of the Ti-6Al-4 V titanium alloy after ion-beam mixing and post-irradiation annealing.

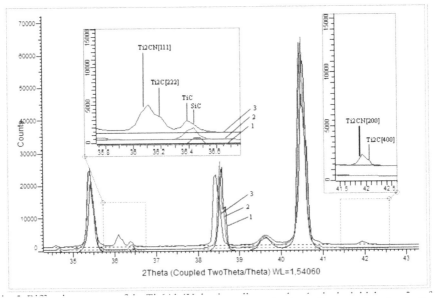

FIGURE 26.5 Diffraction patterns of the Ti-6Al-4 V titanium alloy samples: 1 – in the initial state, 2 – after the ion-beam mixing of Al, 3 – after mixing and heat treatment.

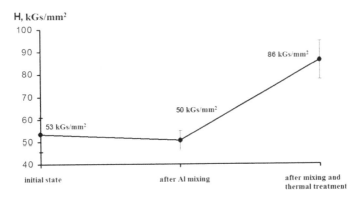

FIGURE 26.6 The micro hardness value of the Ti-6Al-4 V titanium alloy samples in the initial state, after the ion-beam mixings, after mixing and heat treatment.

The measurement of micro hardness has shown that if after ion-beam mixing the value of micro hardness within the spread of values does not differ from the micro hardness of the sample in the initial state and is about ~50 kGs/mm², then after the post-irradiation annealing the value of micro hardness increases by more than 30% up to ~86 kGs/mm².

Thus, as a result of the investigation carried out one can make the following conclusion. Implantation in the periodic-pulse mode with $E = 30$ keV of the metal system: Ti-6Al-4 V titanium alloy with an Al film of ~30 nm thickness deposited on it does not result in the mixing of the Al film and titanium alloy matrix. During the process of the post-irradiation annealing in the vacuum of ~10^{-4} Pa at $T = 900$ °C for 30 min, obviously, the melting of the Al film, "non-bound" with the Ti alloy matrix followed by its evaporation, as the temperature of the Al melting is 660 °C takes place. The Al content in the surface layers decreases down to trace amounts of 1–2 at %. Then during the sample cooling with the furnace from 900 °C down to room temperature, carbon and nitrogen of the residual vacuum atmosphere begin depositing and diffusing into the surface layers, forming Ti_2C, TiC carbides and titanium carbonitride Ti_2CN, the release of which leads to the hardening of the surface layers as well as to the increase of micro hardness by more than 30%.

KEYWORDS

- Heat treatment
- Ion-beam mixing
- Mechanical properties
- Surface hardening
- The composition of the surface layers

REFERENCES

1. Kalin, B. A. (1999). Promising Radiation Technologies in Material Science, Engineering Physics, *1*, 3–10.
2. Terent'ev, V. F. (2004). Cyclic Strength of Modern Metallic Materials Taking into Account the entire Fatique Curve Consideration, Promising Materials, *(5)*, 85–92.
3. Yakovleva, Yu T., & Matokhnyuk, Ye L. (2002). The Effect of Cyclic Loading Rate on the Depth of the One of Plastic Deformation of the BHC-25 Alloy, Problems of Strength, *(2)*, 62–65.
4. Sharkeev, Yu P., Ryabchikov, A. I., Kozlov, E. V., Kurzina, Ya I., Stepanova, I. B., Bozhko, I. A., Kalashnikov, M. P., Fortuna, S. V., & Sivin, D. O. (2004). High Intensity Ion Implantation, the Method of Fine Dispersion Inter Metallic Compounds Formation in the Surface Layers of Metals, News from Higher Educational Institutions, Physics, *9*, 44–52.
5. Kurzina, I. A., Kozlov, E. V., Bozhko, I. A., Kalashnikov, M. P., Fortuna, S. V., Stepanov, I. B., Ryabchikov, A. I., & Sharkeev, Yu P. (2005). Structure-Phase State of Surface Ti Layers Modified Under High Intensity of Ion Implantation, RAN Proceeding, Physics Series, *69(7)*, 1002–1006.
6. Sergeyev, V. P., Sungatulin, A. R., Sergeyev, O. V., & Pushkaryov, G. V. (2006). Nano Hardness and Wear Resistance of High Strength Steels 38ХН3МФА and ШХ-15, Implanted with (Al+B), (Ti+B), Ti Ions, Proceedings of Tomsk Poly Technical University, *309(1)*.
7. Nefedov, V. I. (1984). X-ray Electron Spectroscopy of Chemical compounds, Reference Book, M, Chemistry, 256ps.

CHAPTER 27

INFLUENCE OF PULSE LASER BEAMING WITH VARIOUS DOSES ON SEGREGATION PROCESSES IN SURFACE LAYERS OF CARBON-COATED $Cu_{50}Ni_{50}$ FOILS

A. V. ZHIKHAREV, I. N. KLIMOVA, S. G. BYSTROV,
YA. V. BAYANKIN, E. V. KHARANZHESKY, and V. F. KOBZIEV

CONTENTS

27.1 INTRODUCTION

Diffusion is one of the most important ways of matter transfer in metals. It affects many processes going on in solids. The speed and mechanisms of diffusion are determining factors in the origin of different physical-chemical processes in metals and alloys. Understanding the regularities of diffusion processes will allow obtaining new materials with the changed properties different from those of their initial state [1–3].

The present work is devoted to investigating the effect of the number of laser impulses on the diffusion atoms transfer of the deposited carbon layer into the depth of the matrixes ($Cu_{50}Ni_{50}$ foil), which is in the non-equilibrium elastic-stressed state. Alongside with this the peculiarities of the metal foil interaction with the indentation element as well as the effect of carbon on the change of the micro hardness in the surface layers of the system investigated were studied.

27.2 OBJECTS AND METHODS OF RESEARCH

The objects of investigation were the $Cu_{50}Ni_{50}$ foils rolled to the thickness of 50 μm. After cold longitudinal rolling, the foils were in a strongly non-equilibrium elastically strained state. The concentration of dislocations at such degree of deformation (70,80 %) can reach the values of $(10^9 \div 10^{10})$ sm^{-2} [4, 5]. Further the samples were polished and carbon as graphite was deposited on one side of the sample.

Laser effect on the samples surfaces was produced by the focused beam from the matrix side where carbon had been deposited. The pulse fibro-optical ytterbium laser "Ldesigner F1" (Ateko) was used as a source of irradiation. The area of irradiation was (10×10) mm. Thus the samples were mounted in the beam focus of the laser. The focus distance was 250 mm, with the diameter of the laser focal spot being 36 μm. To treat the entire surface of the sample, the scanning step was set up the same as the diameter of a focal spot. The wavelength of the generated radiation according to the passport of device was 1,064 μm. The power density of the laser irradiation (q_f) applied to the foil surface during the impulse effect was equaled 1.8×10^7 W/cm^2. The number of impulses at a point was taken to be equal 1, 3, 5, 7, 9, 11, 15 and 20 impulses. The impulse repetition rate was 20 kHz, the impulse duration 100 ns. The scanning speed of the laser beam was adjusted so that the given number of impulses got the area equal to the diameter of the focal spot. The irradiation was carried out in the atmosphere of argon.

The element composition of the surface layers of the samples before and after a laser radiation was investigated using x-ray photoelectron spectroscopy (XPS) on the "SPECS" spectrometer. The vacuum in the spectrometer chamber was ~10^{-7} Pa. The x-ray spectra were excited by the MgKα-radiation with an energy of 1253.6 eV. The level-by-level analysis was carried out using surface etching by ions of argon.

The calculated etching rate was ~1 nm/min. A relative error in the determination of the elemental concentration did not exceed 5 % of the measured value.

The analysis of topography of the sample surface before and after a laser irradiation was examined on a scanning probe microscope "Solver P47 Pro" (NT-MDT). The surface scanning was performed using atomic force microscopy (AFM) in the contact mode. As a result of scanning each sample about 10 scan of its irradiated surface were obtained. The size of one scan was (140×140) µm. To obtain the digital parameters values of the roughness of the irradiated area, the recorded scan were processed by the Roughness Analysis technique. For this purpose the special software was used (Image Analysis v.2.1.2, NT-MDT).

The optical images of the surface relief of the foils were obtained with an optical video system of the scanning probe microscope "Solver P47 Pro."

The micro hardness of the samples was measured and calculated in accordance with the GOST (State Standard) 2999–75. The measurements were performed on a "PMT-3" micro hardness meter at a load of 10 g; the exposure under load was 5 s. In order to improve the reliability of the obtained data, the micro hardness test procedure was carried out ten times for each test state of the sample. The results obtained were averaged, and the standard deviation of the micro-hardness values measured was calculated.

27.3 RESULTS AND DISCUSSION

In the AFM-images the surface morphology of the initial $(Cu_{50}Ni_{50})$ +C samples represents a relief of a variable profile with no abrupt changes in heights (Fig. 27.1). Thus of obvious differences in the topography of the surface with carbon and without it was not observed. The AFM-analysis of the roughness parameters revealed that the average height of the surface relief with the deposited carbon ~1.8 µm and the root-mean-square roughness (R_q) ~191.4 nm.

a   *b*

FIGURE 27.1   AFM–Images of the surface topography of the initial $(Cu_{50}Ni_{50})$ +C samples [140×140 µm].

The mean value of the measured surfaces micro hardness (H) of the initial $Cu_{50}Ni_{50}$ foils was ~80 kgs/mm². The micro hardness of the samples surface after carbon deposition increased up to (108.3±4.9) kgf/mm^2 (Fig. 27.2).

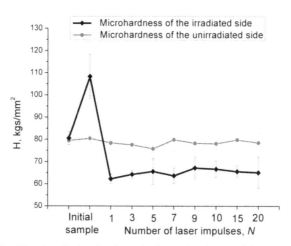

FIGURE 27.2   The plot of micro hardness in the surface layers of the $(Cu_{50}Ni_{50})$ +C samples.

Results of the X-ray photoelectron spectroscopy investigation of the surface layers of the $Cu_{50}Ni_{50}$ foils before carbon deposition indicate that the positions of maxima in $Cu2p_{3/2}$ and $Ni2p_{3/2}$ spectra on all depth of etching do not differ from the binding energy values for pure nickel (E_b = 852.6 eV) and copper (E_b = 932.4 eV). In the near-surface areas (up to 5 nm) in spectra of copper and nickel are present components from the side of higher binding energy (Fig. 27.3). According to the energy position these components [6] can correspond to oxides and hydroxides of copper and nickel that are natural referred to oxygen-containing impurities present on the samples surface.

After carbon deposition on the foils surface the main contribution to XPS-spectra of all analyzed layer is made by the C1 s (E_b = 284.4 eV) line. In the near-surface area, despite of weak intensity of $Cu2p_{3/2}$ line, its asymmetry from side of higher binding energies is still visible, which cannot be said about the $Ni2p_{3/2}$ line, since its intensity is at the level of a background.

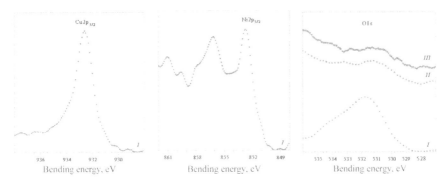

FIGURE 27.3   XPS-spectra of $Cu2p_{3/2}$, $Ni2p_{3/2}$ and O1 s of the initial $Cu_{50}Ni_{50}$ foils.

The results of the effect of the focused laser beam on the samples are the following. The optical images of all irradiated surfaces did not exhibition obvious signs of laser influence (Fig. 27.4). However, according AFM-images it was found out that an increase in number of the laser impulses there is a smoothing the irradiated surface (Fig. 27.5). The Roughness Analysis data showed decreasing the average height of a relief from 1.8 μm in an initial state down to 137.2 nm under irradiation with the maximal number of impulses ($N = 20$). The root-mean-square roughness (R_q) of the surfaces has decreased from 191.4 down to 13.5 nm. On the backside of the samples in all cases of irradiation no changes of the surface relief in comparison with the initial state were observed.

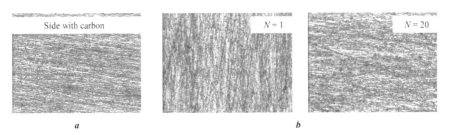

a – initial surface state; b – surface after laser effect by the focused beam

FIGURE 27.4   Optic-images of the surface relief of the irradiation $(Cu_{50}Ni_{50})$ +C samples [×320 (1180×950 μm)].

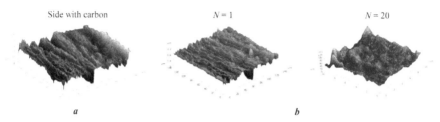

Side with carbon   N = 1   N = 20

a   b

a – initial surface state; *b* – surface after laser effect by the focused beam

FIGURE 27.5 AFM-images of the surface topography of the irradiation $(Cu_{50}Ni_{50})$ +C samples [140×140 µm]

Thus from the AFM-data analysis one can suggest that the irradiated surface samples were exposed to strong enough thermal influence. The rough calculation of temperatures in the area of irradiation of samples showed that the temperature in the field of laser influence could reach values ~(900÷1000)°C. The calculation was carried out by Eq. (1) according to [7, 8].

$$T = \frac{q_f \left(1 - R_{ref}\right) N r_f}{\lambda} \left(1 - \frac{v_{sc} r_f}{4\chi}\right) + T_0 \tag{1}$$

where q_f – power density of the laser irradiation, W/m²; R_{ref} – the coefficient of the reflectivity of the surface of the material; N – number of laser impulses; r_f – radius of the focal spot, m; λ – thermal conductivity of the material, W/ (m*K); v_{sc} – scanning speed [m/s]; χ – thermal diffusivity of the material, m²/s; T_0 – initial temperature of the sample, K.

The analysis of XPS-data showed change of the element composition of the samples surface after laser effect. In comparison with the initial state the content of copper and nickel increased on the irradiated side, with the carbon concentration decreasing (Fig. 27.6). The oxygen concentration in the near-surface layers with the ~5 nm thickness depending on the radiation impulses number decreased by two and more times. Thus lines $Cu2p_{3/2}$ and $Ni2p_{3/2}$ have high intensity and are symmetric practically on all depth of an analyzed layer (Fig. 27.7), which testifies a significant decrease of the O-containing component on the irradiated sides of the foils.

The changes of the elements composition observed in the irradiated surface layer of the material can be accounted for in the following way. Carbon is known to be chemically inert at usual temperatures and at rather high temperatures it combines with many elements revealing strong reduction properties (~800°C and higher) [9]. The rough calculation of the temperatures in the area of irradiation showed that the temperature could reach the values ~(900÷1000)°C. Therefore, in the chosen modes of laser radiation in the surface layer of the samples due the deposited carbon on the foil the reduction of the metals oxides up to pure metals took place. On the un-irradiated side of the foil, no changes in the composition were observed.

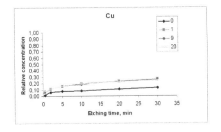
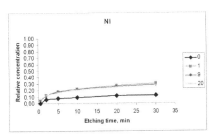
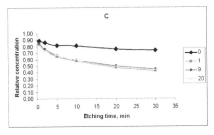

FIGURE 27.6   Elements concentration profiles according to the depth in the surface layers of the $(Cu_{50}Ni_{50})$+C samples.

To confirm the possibility of such reactions, the thermodynamic calculation of temperature at which the reaction of reduction of metals oxides involved in the composition of the surface layers of the system studied will start, has been made [10]. The calculation showed that the reduction of the nickel and copper oxides by carbon and carbon oxide was possible in the range of temperatures $(300 \div 400)° C$ [11, 12].

The effect of the focused laser influence on the micro hardness of the irradiated surface layer in the investigated samples under the chosen modes of radiation revealed itself in the decrease of the micro hardness value on increasing the number of laser impulses. Most probably, it is connected with change of element composition of the irradiated surface layer, which resulted in the enrichment of the analyzed layer (~30 nm) of the irradiated surface with pure copper and nickel, as well as in the decrease of carbon content and complete disappearance of oxygen. No changes of the micro hardness were observed on the backside of the samples.

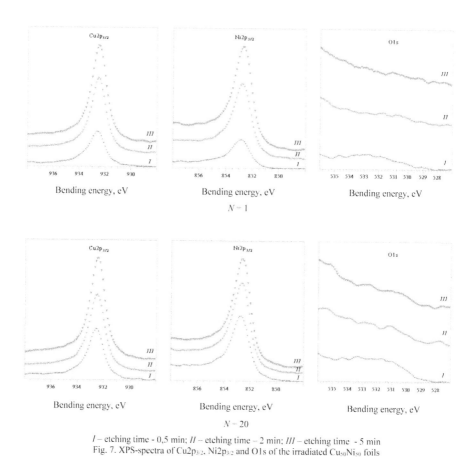

I – etching time - 0,5 min; *II* – etching time – 2 min; *III* – etching time - 5 min
Fig. 7. XPS-spectra of Cu2p$_{3/2}$, Ni2p$_{3/2}$ and O1s of the irradiated Cu$_{50}$Ni$_{50}$ foils

FIGURE 27.7   XPS-spectra of Cu2p$_{3/2}$, Ni2p$_{3/2}$ and O1 s of the irradiated Cu$_{50}$Ni$_{50}$ foils.

27.4   CONCLUSIONS

It has been found out that on increasing the number of laser impulses the intensity of the laser effect on the irradiated surface of the (Cu$_{50}$Ni$_{50}$) +C system also increases. The relief of the samples surface in the area of irradiation at the micro level smoothes and the surface looks melted. The values of temperatures in the area of irradiation calculated for each mode of laser radiation showed that the temperature could reach the values of ~(900÷1000)°C.

The analysis of the chemical composition of the irradiated surface showed that the laser irradiation resulted in the decrease of the carbon concentration and increases of the copper and nickel concentration at all depth of the analyzed layer. The copper and nickel spectra being symmetric at all depth of the layer analyzed, with the O1 s line in the oxygen spectrum being at the background level. It is supposed that carbon assisted the reduction of the metals oxides in to pure metals. The thermodynamic calculation showed the possibility of such reducing reactions taking place.

In all of the cases of laser effect one could observe the decrease of micro hardness in the area of the samples irradiation. It can be attributed to the change of element composition of the irradiated surface layer of the material investigated.

ACKNOWLEDGMENTS

The work was supported by the Russian Foundation for Basic Research (project N 13-02-96002_р_урал) and the programs of the presidium of the Russian Academy of Science (12-П-2-1040 and 12-П-2-1013).

KEYWORDS

- **Carbon**
- **$Cu_{50}Ni_{50}$ alloy**
- **Diffusion**
- **Micro-hardness**
- **Non-equilibrium state**
- **Pulse laser irradiation**
- **Scanning probe microscopy**
- **Segregation**
- **XPS**

REFERENCES

1. Sam, M. F. (1996). Laser and Their Application, Soros Educational Journal, 6, 92–98.
2. Arutyunyan, R. V., Baranov, Yu V., Bol'shov, L. A. et al. (1989). Laser Radiation effect on Materials, Nauka, M, 367p.
3. Rykalin, N. N., Uglov, A. A., Zuev, I. V. & Kokora, A. N. (1985). Laser and Electron beams Treatment of Materials, Mashino Stroyeniye, M, 496p.
4. Polukhin, P. I., Gorelik, S. S, & Vanttsov, V. K. (1982). Physical Basics of Plastic Deformations, M. Metallurgy, 584p.
5. Umansky, Y. S., Skakov, A. N., Ivanov, A. N., Rastorguyev, L. N. (1982). Crystallography, X-ray Diffraction and Electron Microscopy, M., Metallurgy, 632p
6. Nefedov, V. I. (1984). X-Ray Electron Spectroscopy of Chemical compounds, Reference Book, M. Chemistry, 256p.

7. Veiko, V. P. (2005). Lectures Summary on the Course "Physico Technical Basic Knowledge of Laser Technologies, " Section Laser Micro Treatment, SPb Publishing House of Street Petersburg State University ITMO, 110p.

8. Veiko, V. P, Shakhno, Ye A. (2007). Collection of Problems on Laser Technologies, SPb Publishing House of Street Petersburg State University ITMO, 67 p.

9. URL: http://ru.wikipedia.org/wiki/Углерод (date of the reference of 5/7/2013).

10. Krasnov, K. S. (1995). Physical Chemistry, M Higher School, *1*, 512p.

11. URL: http://www.xumuk.ru/inorganic_reactions/search.php?
Query=Ni O+%3D+Ni&sselected=%5D&go.x=21&go.y=19 (Date of the Reference of 5/7/2013.

12. URL: http://www.xumuk.ru/inorganic_reactions/search.php?
Query=Cu O+%3D+Cu&sselected=%5D&go.x=29&>go.y=20 (Date of the Reference of 5/7/2013).

CHAPTER 28

SELF-ORGANIZATION IN PROCESSES UNDER ACTION SUPER SMALL QUANTITIES OF METAL/CARBON NANOCOMPOSITES: REVIEWS ON INVESTIGATION RESULTS

V. I. KODOLOV and V. V. TRINEEVA

CONTENTS

28.1 INTRODUCTION

Earlier in Refs. [1–12], the results of investigations on the obtaining of metal/carbon nanocomposites in polymeric matrixes nanoreactors as well as the influence of these active nanostructures on different media, and also modification of polymeric compositions by super small quantities of metal/carbon nanocomposites are presented. Above investigations are based on such positions of nanochemistry and nanotechnology as self-organization and self-similarity which are fundamentals of fractal theory. Quite often, especially recently, the papers are published, for example, by Malinetsky [13], in which it is considered that nanotechnology is based on self-organization of metastable systems. As assumed [14], self-organization can proceed by dissipative (synergetic) and continual (conservative) mechanisms. At the same time, the system can be arranged due to the formation of new stable ("strengthening") phases or due to the growth provision of the existing basic phase. This phenomenon underlies the arising nanochemistry. Assuming that nanoparticle oscillation energies correlate with their dimensions and comparing this energy with the corresponding region of electromagnetic waves, we can assert that energy action of nanostructures is within the energy region of chemical reactions. Therefore, one of the possible definitions of nanochemistry can be following.

NANOCHEMISTRY IS A SCIENCE INVESTIGATING NANOSTRUCTURES AND NANOSYSTEMS IN METASTABLE ("TRANSITION") STATES AND PROCESSES FLOWING WITH THEM IN NEAR-"TRANSITION" STATE OR IN "TRANSITION" STATE WITH LOW ACTIVATION ENERGIES

To carry out the processes based on the notions of nanochemistry, the directed energy action on the system is required, with the help of chemical particle field as well, for the transition from the prepared near-"transition" state into the process product state (in our case into nanostructures or nanocomposites). The perspective area of nanochemistry is the chemistry in nanoreactors. Nanoreactors can be compared with specific nanostructures representing limited space regions in which chemical particles orientate creating "transition state" prior to the formation of the desired nanoproduct. Nanoreactors have a definite activity, which predetermines the creation of the corresponding product. When nanosized particles are formed in nanoreactors, their shape and dimensions can be the reflection of shape and dimensions of the nanoreactors. The proposed method of corresponding nanostructures synthesis consists in the conducting of redox processes, which proceed in nanoreactors of polymeric matrixes and is accompanied by the reduction of metal ions included into the cavities of organic polymer gels. At the same time, hydrocarbon shells are simultaneously oxidized to carbon. The process realization is possible in liquid phase (the mixing of salts solutions and polymeric solutions) or in solid phase (the joint grinding of metal containing and polymeric phases). However, in both cases the forming metal containing clusters are implanted into inner hollows of polymeric matrixes and orientated on functional groups of these hollows (nanoreactors). It may be said that there is original photography of hollows (nanoreactors) by means of metal containing clusters. Further the transformation process of nanosystems leads to the formation of self-similar and self organized nanostructures represented as metal/carbon nanocomposites which is active to influence on polar media and also on polymeric compositions. Recently, it becomes known that the influence of metal/ carbon nanocomposites super small quantities takes place on the great changes of polymeric materials properties [15–19]. For your attention the review of investigation results on these nanocomposites and the theory of their influence on the polymeric compositions are represented.

28.2   REDOX SYNTHESIS OF METAL/CARBON NANOCOMPOSITES

28.2.1   THEORY OF NANO COMPOSITES FORMATION WITH TIN NANOREACTORS OF POLYMERIC MATRIXES

The synthesis of nanostructures in nanoreactors of polymeric matrixes is the perspective trend for nanochemistry development. Nanoreactors can be compared with specific nanostructures having limited space regions in which chemical particles are orientated creating "transition state" prior to the formation of the desired nanoproduct. Nanoreactors have a definite activity, which predetermines the creation of the corresponding product. When nanosized particles are formed in nanoreactors, their shape and dimensions can be the reflection of shape and dimensions of the nanoreactor, The investigation of redox synthesis of Metal/Carbon nanocomposites in nanoreactors of polymeric matrixes is realized in three stages:

1. The computational designing of nanoreactors filled by metal containing phase and quantum chemical modeling of processes within nanoreactors.
2. The experimental designing and nanoreactors filling by metal containing phase with using two methods (the mixing of salt solution with the solution of functional polymer, for example, polyvinyl alcohol; or the common degeneration of polymeric phase with metal containing phase).
3. The properly redox synthesis of metal/carbon nanocomposites in nanoreactors of polymeric matrixes at narrow temperature intervals.

Previously the authors [20] proposed the parameter called the nanosized interval (B), in which the nanostructures demonstrate their activity. Depending on the structure and composition of nanoreactor internal walls, distance between them, shape and size of nanoreactor, the nanostructures differing in activity are formed. The correlation between surface energy, taking into account the thickness of surface layer, and volume energy was proposed as a measure of the activity of nanostructures, nanoreactors and nanosystems [1]. In this case, we obtain the absolute dimensionless characteristic (a) of nanostructure or nanoreactor activity.

$$a = \varepsilon_s^0 \, d / \varepsilon_v^0 \times N/r(h) = \varepsilon_s^0 \, d / \varepsilon_v^0 \times 1/B, \qquad (1)$$

where $B = r(h)/N$, r – radius of rotating bodies, including the hollow ones, h – film thickness, N – number changing depending on the nanostructure shape. Parameter d characterizes the nanostructure surface layer thickness, and corresponding energies of surface unit and volume unit are defined by the nanostructure composition.

The proposed scheme of obtaining carbon/metal-containing nanostructures in nanoreactors of polymeric matrixes includes the selection of polymeric matrixes containing functional groups. 3d metals (iron, cobalt, nickel, copper) are selected as the elements coordinated on functional groups. The elements indicated easily coordinate with functional groups containing oxygen, nitrogen, halogens. Depending on metal coordinating ability and conditions for nanostructure obtaining (in liquid or solid medium with minimal content of liquid) we obtain "embryos" of future nanostructures of different shapes, dimensions and composition. It is advisable to model coordination processes and further redox processes with the help of quantum chemistry apparatus, following step-by-step consideration in accordance with the planned scheme. At the same time, the metal orientation proceeds in interface regions and nanopores of polymeric phase which conditions further direction of the process to the formation of metal/carbon nanocomposite. In other words, the birth and growth of nanosize structures occur during the process in the same way as known from the macromolecule physics [21], in which Avrami equations are successfully used. The application of Avrami equations to the processes of nanostructure formation was previously discussed in the papers dedicated to the formation of ordered shapes of macromolecules [21], formation of carbon nanostructures by electric arc method [22, 23], obtaining of fiber materials [24].

As follows from Avrami equation –

$$1 - \upsilon = \exp[-k\tau^n], \qquad (2)$$

where υ crystallinity degree, τ duration, k – value corresponding to specific process rate, n – number of degrees of freedom changing from 1 to 6, the factor under the exponential is connected with the process rate with the duration (time) of the process. Under the conditions of the isothermal growth of the ordered system "embryo," it can be accepted that the nanoreactor activity will be proportional to the process rate in relation to the flowing process. Then the share of the product being formed (W) in nanoreactor will be expressed by the following equation.

$$W = 1 - \exp(-a\tau^n) = 1 - \exp[-(\varepsilon_s/\varepsilon_v)\tau^n] = 1 - \exp\{-[(\varepsilon_s^0 d/\varepsilon_v^0)S/V]\,\tau^n\}, \qquad (3)$$

where a nanoreactor activity, ε_s surface energy reflecting the energy of interaction of reagents with nanoreactor walls, ε_v nanoreactor volume energy, $\varepsilon_s^0 d$ multiplication of surface layer energy by its thickness, ε^0_V – energy of nanoreactor volume unit, S – surface of nanoreactor walls, V – nanoreactor volume.

When the metal ion moves inside the nanoreactor with redox interaction of ion (mol) with nanoreactor walls, the balance setting in the pair "metal containing phase–polymeric phase" can apparently be described with the following equation.

$$zF\Delta\varphi = RT\ln K = RT\ln(N_p/N_r) = RT\ln(1-W), \tag{4}$$

where z – number of electrons participating in the process; $\Delta\varphi$ – difference of potentials at the boundary "nanoreactor wall-reactive mixture"; F – Faraday number; R – universal gas constant; T – process temperature; K – process balance constant; N_p – number of moles of the product produced in nanoreactor; N_r – number of moles of reagents or atoms (ions) participating in the process which filled the nanoreactor; W–share of nanoproduct obtained in nanoreactor.

In turn, the share of the transformed components participating in phase interaction can be expressed with the equation, which can be considered as a modified Avrami equation.

$$W = 1 - \exp[-\tau^n\exp(zF\Delta\varphi/RT)], \tag{5}$$

where τ – duration of the process in nanoreactor; n – number of degrees of freedom changing from 1 to 6.

During the redox process connected with the coordination process, the character of chemical bonds changes. Therefore, correlations of wave numbers of the changing chemical bonds can be applied as the characteristic of the nanostructure formation process in nanoreactor.

$$W = 1 - \exp[-\tau^n(n_{HC}/n_{KC})], \tag{6}$$

where n_{HC} corresponds to wave numbers of initial state of chemical bonds, and n_{KC} – wave numbers of chemical bonds changing during the process.

Modified Avrami equations were tested to prognosticate the duration of the processes of obtaining metal/carbon nanofilms in the system "Cu – PVA" at 200°C [25]. The calculated time (2.5 h) correspond to the experimental duration of obtaining carbon nanofilms on copper clusters.

The nanostructures formed in nanoreactors of polymeric matrixes can be presented as oscillators with rather high oscillation frequency. It should be pointed out that according to references [19, 26, 27] for nanostructures (fullerenes and nanotubes) the absorption in the range of wave numbers 1300–1450 cm⁻¹ is indicative. These values of wave numbers correspond to the frequencies in the range 3.9–4.35×10^{13} Hz, that is, in the range of ultrasound frequencies.

If the medium into which the nanostructure is placed blocks its translational or rotational motion giving the possibility only for the oscillatory motion, the nanostructure surface energy can be identified with the oscillatory energy.

$$\varepsilon_s \approx \varepsilon_\kappa = m\upsilon_\kappa^2/2, \tag{7}$$

where m – nanostructure mass, a υ_κ – velocity of nanostructure oscillations. Knowing the nanostructure mass, its specific surface and having identified the surface energy, it is easy to find the velocity of nanostructure oscillations.

$$\upsilon_\kappa = \sqrt{2\varepsilon_\kappa/m} \tag{8}$$

If only the nanostructure oscillations are preserved, it can be logically assumed that the amplitude of nanostructure oscillations should not exceed its linear nanosize, that is, $\lambda < r$. Then the frequency of nanostructure oscillations can be found as follows.

$$\nu_\kappa = \upsilon_\kappa/\lambda \tag{9}$$

Therefore the wave number can be calculated and compared with the experimental results obtained from IR spectra.

28.2.2 QUANTUM CHEMICAL MODELING OF METAL/ CARBON NANOCOMPOSITES FORMATION

28.2.2.1 QUANTUM CHEMICAL CALCULATING EXPERIMENTS FOR PROGNOSIS OF REDOX SYNTHESIS IN POLYMERIC MATRIXES WITH SALT SOLUTION APPLICATION OR METAL OXIDE USING

Usually the probability of self-organization processes increases at the directed action or in the particle flow, based on Prigozhin's theory [28]. Consequently these processes have to take place:

1. At the directed interaction between chemical particles and active centers in micropores;
2. In surface and interface layers of lamellar (polymeric organic and inorganic) systems;
3. On membranes or in intracellular space of biological objects.

The driving force of self-organization processes (formation of nanostructures with definite shapes) is the difference of potentials between the interacting particles and walls of the object that stimulates these interactions. In turn, the directedness of the process between the particles is determined by the interactions between charges and dipoles. At the same time the charge is shifted by external electric and electromagnetic fields of a certain intensity or energy field of the particles being formed those changes at the interaction between the nanoreactor walls and flow of chemical particles being taken inside if the temperature on concentration gradient is available along the nanoreactor. For instance during the adsorption of chemical particles with their following transformation into nanostructures on the corresponding metal substrate templates. Here the flow of electrons excited by the heat energy moves upwards. The fluctuations of heat energy on the template determine the level of transformations of initial compounds into nanostructures on separate sectors of the template [29]. At the electrochemical action onto carbon electrodes [30], in the ion flow carbon particles transform into nanostructures of various shapes similar to those of bodies of revolution.

The formation of nanostructures in xerogels and lamellar substances with energy-saturated channels is slightly different. In such cases the flow of ionized particles is formed due to the difference in the state of charge of the channel walls (nanoreactors) and chemical particles contained in the solution or melt flowing into the channels. When the matrix is placed between electrodes at a small potential and when the solvent is evaporated under vacuum or temperature action, the channels and ion flows in them are oriented and chemical particles of different nature (organic and inorganic) are transformed into nanostructures.

Much more processes have to take place in cathode and anode zones, as well as in inter electrode area of nanoreactors of active lamellar media. In such case, apart from redox processes, it is necessary to take into consideration the exchange processes and interaction reactions of reaction byproducts (sometimes low-molecular) with the walls of nanoreactors. Therefore, it is interesting to investigate the composition, structure and potential being created on the walls of nanoreactors. The potential jump on the boundary "nanoreactor wall-reacting particles" is defined by the charge of the wall surface and the size of the reacting layer that, in turn, depends upon the external energy fed and surface energy (energy stored during the nanoreactor formation). Presumably the energy required to obtain carbon tubules or fullerenes from their "embryos" is less by an order than the formation energy of "embryos" themselves. If redox process is considered the main process preceding the formation of tubules and fullerenes, the work for carrying the charge corresponds to the formation energy of corresponding nanoparticles in the reacting layer. Then the equation of energy conservation for a nanoreactor during the formation of a mol of nanoparticles will be as follows:

$$nF\Delta j = RT \ln N_p / N_r,\qquad(10)$$

where n – number reflecting the charge of chemical particles flowing into the nanoreactor; F – Faraday number; Δj – difference of potentials between the nanoreactor walls and flow of chemical particles; R – gas constant; T – process temperature; N_p – molar share of nanoparticles obtained; N_r – molar share of initial reagents from which the nanoparticles are obtained.

It can be presumed that nanoreactor walls are not inert, but they are participating in the process. In this case the directedness of chemical processes is determined by the peculiarities of adsorption of chemical particles and possibility to form nanophases containing the particles and fragments or separate atoms of nanoreactor walls, as well as the possibility to transport ionized particles and electrons along the nanoreactor. At the same time the thickness of nanoreactor walls can reach the values several times exceeding the distance between walls.

Using the aforesaid equation we can define the values of equilibrium constants when reaching the definite output of nanoparticles, sizes of nanoparticles and with the modification of the equation shapes of nanoparticles being formed. The sizes of internal cavity or reaction zone of the nanoreactor and its geometry significantly influence the sizes and shape of nanoparticles being formed.

The succession of the ongoing processes is conditioned by the composition and parameters (energy and geometry) of nanoreactors. If nanoreactors represent nano-pores or cavities in polymeric gels being formed during the removal of solvents out of gels and their transformation into xerogels or during the formation of crazes in the process of mechanical-chemical treatment of polymers and inorganic phase in the presence of an active medium, the essence of processes in nanoreactors is as follows:

First a nanoreactor is formed in the polymeric matrix that in its geometrical and energy parameters corresponds to the transition state of reagents participating in the reaction; afterwards the nanoreactor is filled with the reaction mass containing reagents and a solvent. The latter is removed and only the reagents oriented in a definite way are left in the nanoreactor. If a sufficient energy impulse is available, for instance, energy isolated during the formation of coordinating bonds between the fragments of reagents and functional groups located in nanoreactor walls, they interact with the formation of the required products. The initially produced coordinating bonds are destroyed simultaneously with a nanoproduct formation.

Thus the main role of nanoreactors is to contribute to the formation of activated complex and the decrease in activation energy of the main reaction between reagents.

To perform such processes it is necessary to preliminary select the polymeric matrix containing nanoreactors (in the form of nanopores or crazes) suitable to the process. Such selection can be carried out with the help of computer chemistry. Further the calculation experiment is made with reagents placed into the nanoreactor with corresponding geometrical and energy parameters. The examples of such calculations in gels of polyvanadium acid and its derivatives, as well as in gels of polyvinyl alcohol are discussed [31, 32]. Experimental confirmation of the correctness of the selection of polymeric matrixes and nanoreactors to obtain carbon-metal-containing nanostructures is reported [33, 34]. Quantum-chemical investigation was carried out with the extraction of the fragments of polymeric chains with the introduction of corresponding metals into the model of ions or oxides. When several types of polymeric and oligomeric substances are used in the model for the formation of the matrix and nanoreactors, the fragments of the corresponding types of substances are applied. The definite result of the calculation experiment is obtained gradually. Any transformations at the initial stages are taken into account at the further ones. Figure 28.1 a –c demonstrates the pictures of the initial calculation stage and the next stage of the interaction between nickel chloride and fragments imitating polyvinyl alcohol and polyethylenepolyamine.

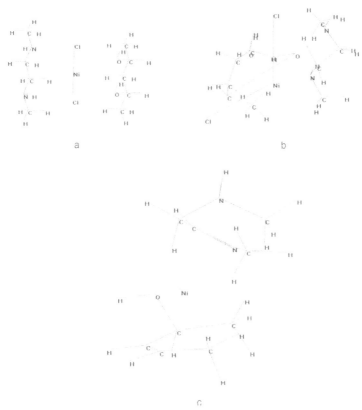

FIGURE 28.1   Stages of the calculation experiment of the interactions between nickel chloride and fragments of PVA-PEPA.

Sizes and shape of nanostructures being formed are determined with the help of the methods of molecular mechanics with a large array of atoms. However, macromolecular systems that are presently accepted as nanostructures were initially registered.

The possibilities of the calculation experiment prognosis:

1.   determination of the possibility to form nanostructures of a definite shape when applying the corresponding metal-containing and polymeric substances;

2. revelation of optimal ratios of metal-containing and polymeric components to obtain the required nanoproduct.

For example, the probable processes at the interaction of PVA with the oxides of 3d-metals are considered in the frameworks of quantum-chemical approximation ZINDO/1 realized in the program product Hyper Chem v.6.03. The evaluation of the interaction possibility of molecule fragments was carried out by the bond length change in them as a result of the geometry optimization, which shows the system interaction and stability. The computational experiment was carried out at the ratios NiO: OH=1:2; 1:4. Figure 28.2 shows the experimental results of the process second stage after the initial processes of dehydration and dehydrogenation. The component ratio NiO: OH = 1: 4.

FIGURE 28.2   Two molecules of 3, 5-dihydroxypentene-1 before the geometry optimization (a), after the geometry optimization with nickel oxide (b).

Thus the energetically more favorable states of the systems are revealed. The scrolling of the macromolecule fragments relatively to the nickel oxide in the shape of hemisphere with the formation of carbon shell is observed in the system. Table 28.1 shows the bond lengths in the molecules before and after the geometry optimization for the systems "3, 5-dihydroxypentene-1-nickel oxide."

TABLE 28.1   Bond Lengths in the Molecules Before and After the Geometry Optimization for the System "3,5-Dihydroxypentene-1-Nickel Oxide" [4]

System	Bond	Bond length, before the optimization, Å	Bond length, after the optimization, Å
Two molecules 3,5-dihydroxypentene-1 and nickel oxide	Ni-O	1.547	2.04565
	C-H	1.09	2.9268

When analyzing the interaction of the system "3,5-dihydroxypentene-1-nickel oxide" it can be concluded that the metal is reduced by polygene, which is formed during the dehydrogenation of PVA as a result of breaking C–H bond and formation of hydrogen radical. Based on the data of theoretical models the ratios of components are selected according to coordination interactions. For nickel, cobalt, copper compounds the molar ratio of the components = 1:4 (metal: number of PVA functional groups). Based on the modeling results obtained it can be expected that the availability of metals of different nature than nickel in the system will result in the distortion of nanostructure shapes obtained, thus influencing the second thermal-chemical stage of nanostructure obtaining.

28.2.2.2 QUANTUM CHEMICAL MODELING FOR THE FUNCTIONALIZATION OF METAL/CARBON NANOCOMPOSITES BY MEANS OF PHOSPHORUS CONTAINING COMPOUNDS [6]

To define the possibility of the actual process of ammonium polyphosphate interaction with nanostructures, the models imitating the functionalization process of carbon nanostructures were built and optimized with the help of software Hyper Chem v. 6.03.

The functional groups were grafted by joint mechanical and chemical treatment of metal/carbon nanocomposite (metal/C NC) and APPh, therefore the interaction of metal/C NC fragment and APPh fragment are considered.

The calculation demonstrates the change in the distance between oxygen atom and nitrogen atom (N^+–O^-) in APPh molecule; it varies in some cases from 1.36 Å to 2.21 Å, which indicates the breaking-off of bonds between oxygen and nitrogen with further isolation of NH_3 from the reaction mass. The distance between phosphorus atom and oxygen atom (P=O) increases, thus confirming the bond transition from (P=O) in ($P^+ \rightarrow O^-$) with further orientation of hydrogen atoms being released from graphene plane surface in relation to oxygen (for bond (O-H) the distance changes from 0.95 Å to 3.02 Å) with the possibility of (P) OH group formation. The change in the configuration of molecule models indicates the progress of interactions.

Thus quantum-chemical modeling allows assuming the possibility of polyphosphate practical application for nanocomposite modification.

28.2.2.3 QUANTUM CHEMICAL INVESTIGATION OF NITROGEN CONTAINING FRAGMENTS INTERACTION WITH METAL/ CARBONIC NANOCOMPOSITES [6]

In the process of quantum-chemical modeling the fragments imitating polyethylene polyamine (PEPA), cobalt/carbon nanocomposite (Co/C NC), nickel/carbon nanocomposite (Ni/C NC), copper/carbon nanocomposite (Cu/C NC) with the inclusion of Co^{2+}, Ni^{2+}, Cu^{2+} ions are optimized (Fig. 28.3).

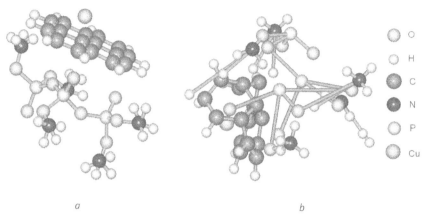

a b

FIGURE 28.3 Modeling of functionalization process on the example of copper/carbon nanocomposite (a – before geometry optimization, b – after optimization).

As the modification process initially assumed the production of fine suspensions (FS) of NC on PEPA basis, the fragments imitating the behavior of the corresponding suspensions of Co/C, Ni/C, Cu/C nanocomposites were optimized (Fig. 28.4) and their absolute values of binding energy $E_{PEPA-NC(Co)}$, $E_{PEPA-NC(Ni)}$, $E_{PEPA-NC(Cu)}$ were defined.

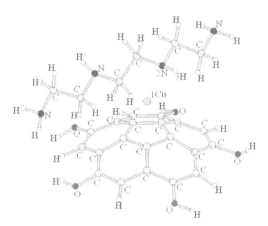

FIGURE 28.4 Fragment of Cu/C NC FS on PEPA basis.

The next step was to model the influence of fine suspensions of nanocomposites on epoxy resin. The complexes formed similarly with the previous ones were optimized, and the absolute values of binding energy were found for each of them $E_{EDR-PEPA-NC(Co)}$, $E_{EDR-PEPA-NC(Ni)}$, $E_{EDR-PEPA-NC(Cu)}$. The absolute values of binding energy are given in Table 28.2.

TABLE 28.2 Absolute Binding Energies of the Fragments

	$E_{NC(Me)}$, KJ/mol	$E_{PEPA-NC(Me)}$, KJ/mol	$E_{EDR-PEPA-NC(Me)}$, KJ/mol
Co^{2+}	−19116.50	-29992.05	-51486.96
Ni^{2+}	−18562.38	-29098.04	-50621.15
Cu^{2+}	−18340.32	-28764.72	-50315.94
E_{EDR}, KJ/mol	−21424.25		
E_{PEPA}, KJ/mol	−10131.36		

Using the data [5] by the following formula:

$$E_1 = E_{EDR-PEPA-NC(Me)} - E_2 - E_{EDR}$$
$$Å_2 = Å_{PEPA-NC(PEPA)} - Å$$

(11)

the relative interaction energies of molecular complexes E_1 are calculated and the diagram is arranged (Fig. 28.5).

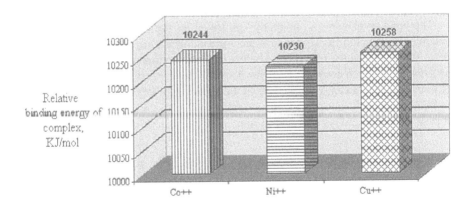

FIGURE 28.5 Diagram of relative interaction energies of molecular complexes.

From the diagram it is seen that the relative interaction energy of molecular complexes with Cu/C NC is higher in comparison with the complexes with Co/C NC and Ni/C NC content. As the polymer forms the strongest complexes with Cu/C

NC, therefore after the modification it will be the most effective. The detailed analysis of the lengths of the bonds formed and effective charges before and after the optimization of fragments imitating PEPA interaction with Cu/C NC (Tables 28.3 and 28.4) indicates that stable coordination bonds were formed between NC fragment and PEPA (between copper ion and nitrogen atom of amine group NH of PEPA). It was found that after the interaction of two fragments studied a part of electron density of N atom participating in the bond shifted to Cu atom, thus resulting in NH bond weakening.

TABLE 28.3   Bond Lengths

Bond designation	Bond length before optimization, Å	Bond length after optimization, Å
Cu-N	2.82	1.95
Cu-C	2.65	2.25

TABLE 28.4   Effective Charges

Atom number (see Fig. 28.2)	Atom designation	Effective charge before optimization	Effective charge after optimization
1	Cu (copper)	0.138	−0.250
2	N (nitrogen)	−0.066	0.420
3	H (hydrogen)	0.045	0.027

The bond weakening is indirectly confirmed by the increase in the effective charge of H atom and slight change in the wave number in oscillatory spectra calculated. The wave number of NH bond before the optimization was 3360 cm^{-1}, and after 3354 cm^{-1}, which correlates with the data of IR spectra obtained with the help of IR Fourier spectrometer. For instance, in the spectrum of PEPA and NC suspension on PEPA basis with nanocomposite concentration 0.03% the shift of wave numbers of peaks of amine groups is observed from 3280 cm^{-1} to 3276 cm^{-1}.

28.2.3   EXPERIMENTAL INVESTIGATIONS OF REDOX SYNTHESIS OF METAL/CARBON NANOCOMPOSITES [5]

For the synthesis of nanoparticles and nanowires from the mixture of metal salts and polyvinyl alcohol (PVA), the aqueous solutions of salts were mixed in a certain ratio with the aqueous solution of PVA. The mean molar ratio of PVA in the mixture was 5. The experiments were carried out on the glass substrates; after the obtained mixtures had been dried, they formed colored transparent films. On some samples, the films were broken due to a large surface tension. The films were heated at $t = 250°C$

until their color, composition and morphology changed. For the control over the process, a complex of methods was used, i.e. photocalorimetry, optical microscopy, X-ray photoelectron spectroscopy and atomic power microscopy.

When PVA was added to the powders of metal chlorides, the color of the mixture changed: the mixture of copper chloride became yellow-green, the cobalt chloride mixture – blue, and nickel chloride – pale-green (Fig. 28.6).

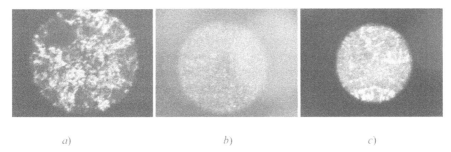

a)   *b)*   *c)*

FIGURE 28.6   The photographs of the samples containing PVA and copper chloride (*a*), cobalt chloride (*b*), and nickel chloride (*c*).

Observing the color changes, one can draw a conclusion that when PVA interacts with metal chlorides, the formation of complex compounds takes place.

Among the above-discussed metals, iron is most active. Brown-red inclusions on the photograph evidence the formation of the complex iron compounds. In addition, on all the photographs depicting the mixtures containing metal chlorides, one can see a net of weaves, which are most likely the reflections of nanostructures.

In order to compare these structures, the investigations of the morphology of the films changing over a certain range of temperatures were carried out with the help of atomic power microscopy (Fig. 28.7).

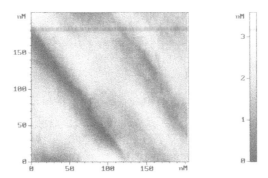

FIGURE 28.7   The Micrographs of the surface geometry of the PVA film with nickel chloride.

When the nanoproduct pictures obtained by atomic power microscopy and optical microscopy are compared with the TEM micrograph of the nanoproduct treated thermally and with aqueous solution for the matrix removal, one can notice some correspondence between them. The nanoproduct represents interweaving tubulens containing Cu(I), Cu(II). In Fig. 28.6, there are also optical effects indicating light polarization at light transmission through the films owing to the defects appearing during the formation of the complex compounds at the initial stage of the process.

Due to the fact, that metal ions are active, in the polymer medium they immediately appear in the environment of the PVA molecules and form bonds with the hydroxyl groups of this polymer. Polyvinyl alcohol replicates the structure of the particle that it surrounds; however, due to the tendency of the molecules of the metal salts or other metal compounds to combine, PVA as if envelops the powder particles, and therefore the forms of the obtained nanostructures can be different. The optical microscopy method allows to determine the structure of the nanostructures at the early stage.

The methods of optical spectroscopy and X-ray photoelectron spectroscopy allows to determine the energy of the interaction of the chemical particles in the nanoreactors with the active centers of the nanoreactor walls, which stimulate reduction-oxidation processes. Depending on the nature of the metal salt and the electrochemical potential of the metal, different metal reduction nanoproducts in the carbon shells differing in shape are formed. Based on this result we may speak about new scientific branch nanometallurgy. The stages of nanostructures synthesis may be represented [35] by the following scheme (Fig. 28.8).

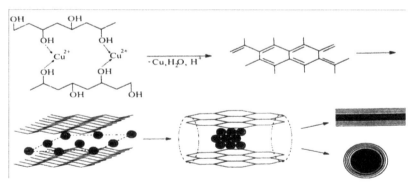

FIGURE 28.8 Schemes of copper carbon nanostructures obtaining from copper ions and Polyvinyl Alcohol.

The possible ways for obtaining metallic nanostructures in carbon shells have been determined. The investigation results allow to speak about the possibility of the isolation of metallic and metal-containing nano-particles in the carbon shells differing in shape and structure. However, there are still problems related to the calculation and experiment because using the existing investigation methods it is difficult unambiguously to estimate the geometry and energy parameters of nanoreactors under the condition of 'erosion' of their walls during the formation of metallic nanostructures in them.

The essence of the method [27] consists in coordination interaction of functional groups of polymer and compounds of 3d metals as a result of grinding of metal-containing and polymer phases (Fig. 28.9). Further, the composition obtained undergoes thermolysis following the temperature mode set with the help of thermo gravimetric and differential thermal analyses. At the same time, we observe the polymer carbonization, partial or complete reduction of metal compounds and structuring of carbon material in the form of nanostructures with different shapes and sizes.

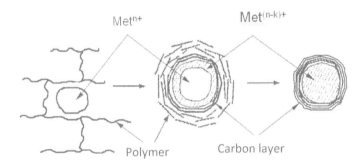

FIGURE 28.9   Mechanisms of nanostructure formation in nanoreactors of polymer matrixes.

Presented mechanism is confirmed by experimental data (microphotograph of transmission electron microscopy). In this case nanofilms are scrolled as seen in Fig. 28.10.

FIGURE 28.10   Microphotograph (transmission electron microscopy) demonstrating the moment of nanofilm scrolling on metal nanoparticles.

Metal/carbon nanocomposite (Me/C) represents metal nanoparticles stabilized in carbon nanofilm structures. In turn, nanofilm structures are formed with carbon amorphous nanofibers associated with metal containing phase. As a result of stabilization and association of metal nanoparticles with carbon phase, the metal chemically active particles are stable in the air and during heating as the strong complex of metal nanoparticles with carbon material matrix is formed.

Below the microphotographs of transmission electron microscopy specific for different types of metal/carbon nanocomposites are demonstrated (Fig. 28.11).

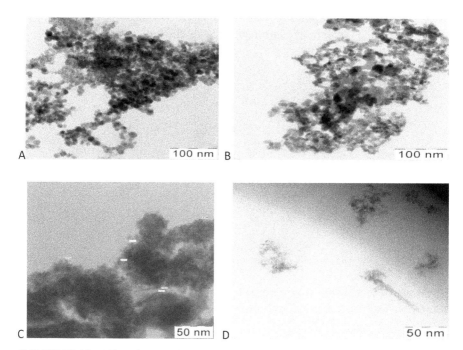

FIGURE 28.11   Microphotographs of metal/carbon nanocomposites: A-Cu/C; B–Ni/C; C–Co/C; D–Fe/C.

One of the main properties of metal/carbon nanocomposites obtained is the ability to form fine suspensions [5–7] in various media (organic solvents, water, solutions of surface-active substances). The average size of nanoparticles in fine suspensions is 10–25 nm depending on the type of metal/carbon nanocomposite (Fig. 28.12).

Distribution of particles by size in organic

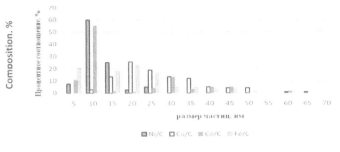

Size of particles, nm

FIGURE 28.12 Distribution of particles by size in organic medium depending on the types of Metal/Carbon nanocomposite.

The short information about nanostructures formation mechanism in polymeric matrix nanoreactors as well as about the methods of synthesis and control during metal/carbon nanocomposites production represents. The main attention is given for the ability of nanocomposites obtained to form the fine dispersed suspensions in different media and for the distribution of nanoparticles in media. The examples of improving technical characteristics of foam concretes and glue compositions are given.

28.2.3.1 NANOREACTORS DESIGNING PROBLEMS AND THE METHODS OF FORMATION OF NANOREACTORS FILLED BY METAL CONTAINING PHASE

The synthesis of Carbon or Metal/Carbon nanostructures usually proceeds with redox reactions, in which hydrocarbon part of reactive mass is oxidized and metal containing phase part partly or almost completely is reduced [1]. The synthesis of nanostructures in nanoreactors of polymeric matrixes Is represented the perspective trend for nanochemistry development. Nanoreactors can be compared with specific nanostructures representing limited space regions in which chemical particles orientate creating "transition state" prior to the formation of the desired nanoproduct. Nanorectors have a definite activity, which predetermines the creation of the corresponding product. When nanosized particles are formed in nanoreactors, their shape and dimensions can be the reflection of shape and dimensions of the nanoreactor [9].

The formation of nanostructures or metal/carbon nanocomposites in polymeric matrixes depends on the nature of metal containing phase and the nature of polymeric matrix, and also the conditions of formation of nanoreactors filled by secondary metal containing phase, as well as the conditions of redox synthesis.

In the case, when the interaction between metal salt solution and polymer solution takes place, metal ion, according to scheme, interacts with the functional groups of macromolecules in the interlayer space or with the functional groups of individual macromolecule. In this time the embryos of metal or metal containing clusters are formed. The sizes and forms of cluster embryos are determined by the metal nature. The clusters obtained associate with the macromolecules oriented inside nanoreactor walls. The creation of different nanosized nanoreactors and nanostructures is possible in dependence on the concentration of polymer solution.

In the case, when the interaction of solid metal containing phase (metal oxides) with polymeric phase in active medium at the intensive grinding occurs, the metal containing clusters formed get in active zones (pores, interlayer space) of polymeric phase and interact with the functional groups of polymeric matrixes. Sizes and forms of nanoreactors created depend on the sizes of active zones in polymeric matrixes. For instance, the nanoreactors obtained inside polyvinyl alcohol matrix with large pores distinguish the great distribution on sizes and forms. Metal/Carbon nanocomposites obtained in these nanoreactors have middle activity.

The investigation of redox synthesis of Metal/Carbon nanocomposites in nanoreactors of polymeric matrixes is realized in three stages:

1. the computational designing of nanoreactors filled by metal containing phase and quantum chemical modeling of processes within nanoreactors.
2. the experimental designing and nanorectors filling by metal containing phase with using two methods:
   a. the mixing of salt solution with the solution of functional polymer (polyvinyl alcohol, polyvinyl chloride, polyvinyl acetate).
   b. the common degenaration of polymeric phase with metal containing phase.
3. the properly redox synthesis of Metal/Carbon nanocomposites in nanoreactors of polymeric matrixes at narrow temperature intervals.

The first and second stages concern to preperatiory stages. On the second stage the functional groups in nanoreactor walls participate in coordination reactions between metal ions (2.a method) or clusters of metal containing phase (2.b method).

The computational experiment was carried out with software products Games and Hyper Chem, with visualization. The definite result of the computational experiment is obtained stage by stage. Any transformations at the initial stages are taken into account at further stages. The prognostic possibilities of the computational experiment consist in defining the probability of the formation of nanostructures of definite shapes when using the corresponding metal containing and polymeric substances, when studying the character of interaction of metal ion, atom, cluster or its compound, fragment of metal containing phase with fragments of nanoreactor walls. The optimal dimensions and shape of internal cavity of nanoreactors, optimal correlation between metal containing and polymeric components for obtaining the necessary nanoproducts are found with the help of quantum-chemical modeling.

The availability of d metal in polymeric matrix results, in accordance with modeling results, in its regular distribution in the matrix and self-organization of the matrix.

For the corresponding correlations "polymer metal containing phase" the dimensions, shape and energy characteristics of nanoreactors are found with the help of AFM [27, 36]. Depending on a metal participating in coordination, the structure and relief of xerogel surface change. The comparison of phase contrast pictures on the corresponding films indicates greater concentration of the extended polar structures in the films containing copper, in comparison with the films containing nickel and cobalt (Fig. 28.13).

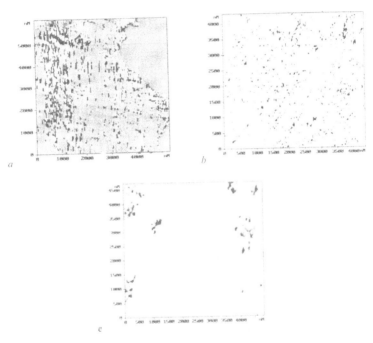

FIGURE 28.13 Pictures of phase contrast of PVA surfaces containing copper (a), nickel (b) and cobalt (c).

The processing of the pictures of phase contrast to reveal the regions of energy interaction of cantilever with the surface in comparison with the background produces practically similar result with optical transmission microscopy. Corresponding to data of AFM the sizes of nanoreactors obtained from solutions of metal chlorides and the mixture of polyvinyl alcohol (PVA) with polyethylene polyamine (PEPA) are determined (Table 28.5).

TABLE 28.5 Sizes of Nanoreactors Found With the Help of Atomic Force Microscopy

Composition	Sizes of AFM formations				
	Length	Width	Height	Area	Density
PVA:PEPA: CoCl2= 2:1:1	400–800	150–400	30–40	60–350	5.5
PVA: PEPA: NiCl2=2:1:1	80–100	80–100	25–35	6–12	120
PVA: PEPA: CuCl2=2:1:1	80–100	80–100	20–30	6–20	20
PVA: PEPA: CoCl2=2:2:1	600–900	300–600	100–120	180–500	3.0
PVA: PEPA: NiCl2=2:2:1	40–80	40–60	10–30	2–4	350

The results of AFM investigations of xerogels films obtained from metal oxides and PVA [15] is distinguished in comparison with previous data that testify to difference in reactivity of metal chlorides and metal oxides.

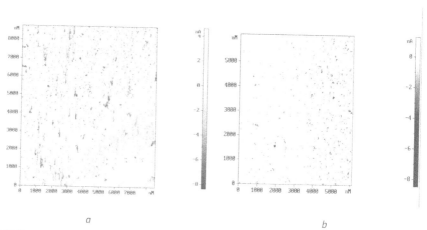

a   *b*

FIGURE 28.14 Phase contrast pictures of xerogels films PVA–Ni (*a*) and PVA–Cu (*b*)

Below the fields of energetic interaction of cantilever with surface of xerogels PVA–Ni (*a*) and PVA–Cu (*b*) are given.

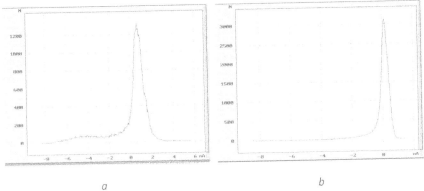

a   b

FIGURE 28.15 The pictures of energetic interaction of cantilever with surface of xerogels

According to AFM results investigation the addition of Ni/C nanocomposite in PVA leads to more strong coordination in comparison with analogous addition Cu/C nanocomposite.

The mechanism of formation of nanoreactors filled with metals was found with the help of IR spectroscopy.

Thus, at the second stage the coordination of metal containing phase and corresponding orientation in nanoreactor take place.

The proposed scheme of obtaining carbon/metal-containing nanostructures in nanoreactors of polymeric matrixes includes the selection of polymeric matrixes containing functional groups. 3d metals (iron, cobalt, nickel, copper) are selected as the elements coordinating functional groups. The elements indicated easily coordinate with functional groups containing oxygen, nitrogen, halogens. Depending on metal coordinating ability and conditions for nanostructure obtaining (in liquid or solid medium with minimal content of liquid) we obtain "embryos" of future nanostructures of different shapes, dimensions and composition. It is advisable to model coordination processes and further redox processes with the help of quantum chemistry apparatus.

The computational experiment was carried out with software products Games and Hyper Chem, with visualization. The definite result of the computational experiment is obtained stage by stage.

Any transformations at the initial stages are taken into account at further stages.

The prognostic possibilities of the computational experiment consist in defining the probability of the formation of nanostructures of definite shapes when using the corresponding metal containing and polymeric substances, when studying the character of interaction of metal ion, atom, cluster or its compound, fragment of metal containing phase with fragments of nanoreactor walls.

The ultimate breaking stresses were compared in the process of compression of foam concretes modified with copper/carbon nanocomposites obtained in different nanoreactors of polyvinyl alcohol.

The sizes of nanoreactors change depending on the crystallinity and correlation of acetate and hydroxyl groups in PVA, which results in the change of sizes and activity of nanocomposites obtained in nanoreactors.

It is observed that the sizes of nanocomposites obtained in nanoreactors of PVA matrixes 16/1(ros) (NC2), PVA 16/1 (imp) (NC1), PVA 98/10 (NC3), correlate as NC3 > NC2 > NC1. The smaller the nanoparticle size and greater its activity, the less amount of nanostructures is required for self-organization effect.

28.2.3.2   THE CONDITIONS OF REDOX SYNTHESIS OF METAL/ CARBON NANOCOMPOSITES AND THE NANOCOMPOSITES CHARACTERIZATION

At the third stage it is required to give the corresponding energy impulse to transfer the "transition state" formed into carbon/metal nanocomposite of definite size and shape. To define the temperature ranges in which the structuring takes place, DTA-TG investigation is applied.

It is known that small changes of weight loss (TG curve) at invariable exo-thermal effect (DTA curve) testify to the self-organization (structural formation) in system. Below typical curves of DTA-TG investigation lead for prognosis of redox synthesis of metal/carbon nanocomposites (Fig. 28.16).

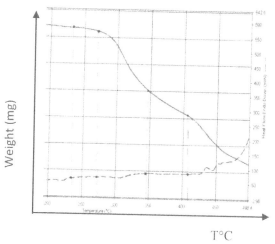

FIGURE 28.16   Typical curves of DTA-TG investigations of xerogels films containing metal and polymeric phases.

According to data of DTA-TG investigation (Fig. 28.16) optimal temperature field for film nanostructure obtaining is 230–270°C, and for spatial nanostructure obtaining 325–410°C.

It is found that in the temperature range under 200°C nanofilms, from carbon fibers associated with metal phase as well (Fig. 28.17), are formed on metal or metal oxide clusters. When the temperature elevates up to 400°C, 3D nanostructures are formed with different shapes depending on coordinating ability of the metal.

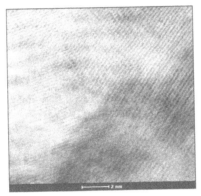

FIGURE 28.17 Microphotographs obtained with the help of transmission electron microscopy. Cu/C nanocomposite.

To investigate the processes at the second stage of obtaining metal/carbon nano-composites X-ray photoelectron spectroscopy, transmission electron microscopy and IR spectroscopy are applied. The sample for IR spectroscopy was prepared when mixing metal/carbon nanocomposite powder with 1 drop of Vaseline oil in agate mortar to obtain a homogeneous paste with further investigation of the paste obtained on the appropriate instrument. As the Vaseline oil was applied when the spectra were taken, we can expect strong bands in the range 2750–2950 cm^{-1}. Two types of nanocomposites rather widely applied during the modification of various polymeric materials were investigated. These were: copper/carbon nanocomposite and nickel/carbon nanocomposite specified below. In turn, the nanopowders obtained were tested with the help of high-resolution transmission electron microscopy, electron micro diffraction, laser analyzer, X-ray photoelectron spectroscopy and IR spectroscopy.

The method of metal/carbon nanocomposite synthesis applied has the following advantages:

1. The perspectives of this investigation are looked through in an opportunity of thin regulation of processes and the entering of corrective amendments during processes.

2. Wide application of independent modern experimental and theoretical analysis methods to control the technological process.
3. Technology developed allows synthesizing a wide range of metal/carbon nanocomposites depending on the process conditions.
4. Process does not require the use of inert or reduction atmospheres and specially prepared catalysts.
5. Method of obtaining metal/carbon nanocomposites allows applying secondary raw materials.

In this investigation the possibilities of developing new ideas about self-organization processes during redox synthesis within nanoreactors of polymeric matrixes as well as about nanostructures and nanosystems are discussed on the example of metal/carbon nanocomposites. It is proposed to consider the obtaining of metal/carbon nanocomposites in nanoreactors of polymeric matrixes as self-organization process similar to the formation of ordered phases. The perspectives of this investigation are looked through in an opportunity of thin regulation of processes and the entering of corrective amendments during processes.

28.3   ENERGETIC CHARACTERISTICS OF METAL/CARBON NANOCOMPOSITES

28.3.1   RELATIONS BETWEEN ENERGETICS AND MORPHOLOGY FOR METAL/CARBON NANOCOMPOSITES

Metal/carbon nanocomposite (Me/C) represents metal nanoparticles stabilized in carbon nanofilm structures [12–14]. In turn, nanofilm structures are formed with carbon amorphous nanofibers associated with metal containing phase. As a result of stabilization and association of metal nanoparticles with carbon phase, the metal chemically active particles are stable in the air and during heating as the strong complex of metal nanoparticles with carbon material matrix is formed.

The test results of nanocomposites obtained are given in Table 28.6.

TABLE 28.6   Characteristic of Metal/Carbon Nanocomposites (Met/C HK)

Type of Met/C NC	Cu/C	Ni/C	Co/C	Fe/C
Composition, Metal/Carbon [%]	50/50	60/40	65/35	70/30
Density, [g/cm³]	1.71	2.17	1.61	2.1
Average dimension, [nm]	20(25)	11	15	17
Specific surface, [m²/g]	160 (average)	251	209	168
Metal nanoparticle shape	Close to spherical, there are dodecahedrons	There are spheres and rods	Nanocrystals	Close to spherical

TABLE 28.6   *(Continued)*

Type of Met/C NC	Cu/C	Ni/C	Co/C	Fe/C
Caron phase shape (shell)	Nanofibers associated with metal phase forming nano-coatings	Nanofilms scrolled in nanotubes	Nanofilms associated with nanocrystals of metal containing phase	Nanofilms forming nanobeads with metal containing phase
Atomic magnetic moment [8], [μB]	0.0	0.6	1.7	2.2
Atomic magnetic moment (nano-composite), [μB]	0.6	1.8	2.5	2.5

To investigates the processes, optical transmission microscopy, spectral photometry, IR.

Under metal/carbon nanocomposite we understand the nanostructure containing metal clusters stabilized in carbon nanofilm structures. The carbon phase can be in the form of film structures or fibers. The metal particles are associated with carbon phase. The metal nanoparticles in the composite basically have the shapes close to spherical or cylindrical ones. Due to the stabilization and association of metal nanoparticles with carbon phase, chemically active metal particles are stable in air and during heating as the strong complex of metal nanoparticles with the matrix of carbon material are formed. The nanocomposites described above were investigated with the help of IR spectroscopy by the technique indicated above. In this chapter, the IR spectra of Cu/C and Ni/C nanocomposites are represented in Figs. 28.18 and 28.19, which find a wider application as the material modifiers.

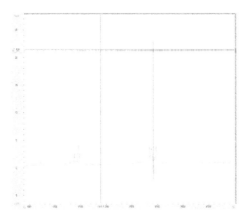

FIGURE 28.18   IR spectra of copper/carbon nanocomposite powder.

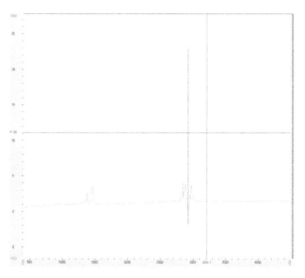

FIGURE 28.19   IR spectra of nickel/carbon nanocomposite powder.

On IR spectra of two nanocomposites the common regions of IR radiation absorption are registered (Figs. 28.18 and 28.19). Further, the bands appearing in the spectra and having the largest relative area were evaluated. We can see the difference in the intensity and number of absorption bands in the range 1300–1460 cm^{-1}, which confirms the different structures of composites. In the range 600–800 cm^{-1} the bands with a very weak intensity are seen, which can be referred to the oscillations of double bonds (π-electrons) coordinated with metals. In case of Cu/C nanocomposite a weak absorption is found at 720 cm^{-1}. In case of Ni/C nanocomposite, except for this absorption, the absorption at 620 cm^{-1} is also observed.

In IR spectrum of copper/carbon nanocomposite two bands with a high relative area are found:

at 1323 cm^{-1} (relative area – 9.28).

at 1406 cm^{-1} (relative area – 25.18).

These bands can be referred to skeleton oscillations of polyarylene rings.

In IR spectrum of nickel/carbon nanocomposite the band mostly appears at 1406 cm^{-1} (relative area–14.47).

According to the investigations with transmission electron microscopy the formation of carbon nanofilm structures consisting of carbon threads is characteristic for copper/carbon nanocomposite. In contrast, carbon fiber structures, including nanotubes, are formed in nickel/carbon nanocomposite. There are several absorption bands in the range 2800–3050 cm^{-1}, which are attributed to valence oscillations of C-H bonds in aromatic and aliphatic compounds. These absorption bonds are connected with the presence of vaselene oil in the sample. It is difficult to find the presence of metal in the composite as the metal is stabilized in carbon nanostructure.

At the same time, it should be pointed out that apparently nanocomposites influence the structure of vaselene oil in different ways. The intensities and number of bands for Cu/C and Ni/C nanocomposites are different:

1. for copper/carbon nanocomposite in the indicated range 5 bands, and total intensity corresponds by the relative area 64.63.
2. for nickel/carbon nanocomposite in the same range 4 bands with total intensity (relative area) 85.6.

The distribution of nanoparticles in water, alcohol and water-alcohol suspensions prepared based on the above technique are determined with the help of laser analyzer. In Figs. 28.20 and 28.21, you can see distributions of copper/carbon nanocomposite in the media different polarity and dielectric penetration. When comparing the figures we can see that ultrasound dispergation of one and the same nanocomposite in media different by polarity results in the changes of distribution of its particles. In water solution the average size of Cu/C nanocomposite equals 20 nm, and in alcohol medium greater by 5 nm.

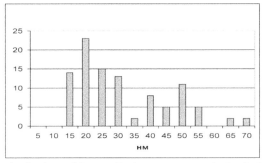

FIGURE 28.20   Distributions of Copper/Carbon nanocomposite in alcohol.

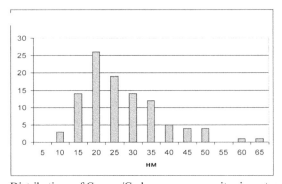

FIGURE 28.21   Distributions of Copper/Carbon nanocomposites in water.

Assuming that the nanocomposites obtained can be considered as oscillators transferring their oscillations onto the medium molecules, we can determine to what extent the IR spectrum of liquid medium will change.

In the paper the possibilities of developing new ideas about self-organization processes during redox synthesis within nanoreactors of polymeric matrixes as well as about nanostructures and nanosystems are discussed on the example of metal/carbon nanocomposites. It is proposed to consider the obtaining of metal/carbon nanocomposites in nanoreactors of polymeric matrixes as self-organization process similar to the formation of ordered phases.

The perspectives of this investigation are looked through in an opportunity of thin regulation of processes and the entering of corrective amendments during processes.

Based on the results of computational experiments the following polymeric matrixes were selected: polyvinyl alcohol (PVA), polyvinyl chloride (PVC), polyvinyl acetate (PVAc), mixture of PVA with polyethylene polyamine (PEPA), derivatives of polyvanadium acid and compounds of such metals as iron, cobalt, nickel and copper. At the same time, after the mixing and sometimes during mechanical and chemical processing the gels saturated with the corresponding metals were produced. During the stage-by-stage control of the processes in nanoreactors the following set of methods was applied: spectrophotometry; optical transmission microscopy; IR spectroscopy; X-ray photoelectron spectroscopy (XPES); atomic force microscopy (AFM); DTA-TG investigation.

At the same time, the directedness of the processes of nanostructure formation, shape and sizes of nanoreactors, as well as future nanostructures, temperature mode of thermal-chemical treatment of xerogels are found. The nanoreactors sizes are found by atomic force microscopy and depend on the nature of polymeric and metal containing phases [26].

Temperature intervals to obtain the nanostructures of different shapes are found by DTA-TG data:

1. obtaining of nanofilms in the temperature range under 200°C;
2. obtaining of globular or cylindrical nanostructures in the temperature range under 400°C.

Temperature ranges found with DTA-TG investigations differ for different mixtures (metal compound–polymer).

At 200°C lamellar nanofilms under 5 nm thick with metal containing nanocrystals between the layers are usually formed. The nanostructures obtained can be referred to as metal/carbon or, in case of synthesis in PVA-PEPA nanoreactors, metal/carbon polymeric nanocomposites. When the temperature elevates up to 400°C, 3D nanostructures are formed with different shapes depending on coordinating ability of the metal. A metal/carbon nanocomposite represents metal nanoparticles stabilized in carbon nanofilmed structures. In these samples the carbon phase can be in the form of film structures or fibers. The metal particles are associated with the carbon phase. Metal nanoparticles in the composite with the size range 7–25 nm (the size is determined by the metal nature) are mainly spherical.

28.3.2 METAL/CARBON NANOCOMPOSITES AS GENERATORS OF ELECTROMAGNETIC WAVES

The metal/carbon nanocomposites are considered as super molecules. Therefore, their surface energies analogously energy of usual molecules consists portions of energy which correspond to progressive, rotation and vibration motions and also electronic motion. From the metal/carbon nanocomposites Raman and IR spectra analysis it follows that the skeleton vibration of them on the vibrations frequencies corresponds to ultrasonic vibrations. The nanocomposite vibrations energy values are determined by the corresponding nanoparticles sizes and masses. Usually metal/carbon nanocomposites have the great dipole moment. Therefore, it is possible the proposition that nanocomposite is vibrator which radiates electromagnetic waves. This hypothesis is confirmed by the increasing of IR spectra lines for liquid media in which the super small quantities of metal/carbon nanocomposites are introduced. For example, IR spectroscopic investigation of fine suspension based on isomethyltetrahydrophthalic anhydride containing 0.001% of Cu/C nanocomposite indicates the decrease in the peak intensity, which sharply increased on the third day when nanocomposite was introduced (Fig. 28.22).

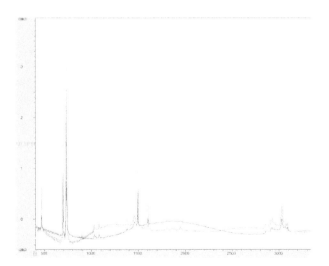

FIGURE 28.22 Changes in IR spectrum of copper/carbon nanocomposite fine suspension based on isomethyltetrahydrophthalic anhydride with time (a–IR spectrum on the first day after the nanocomposite was introduced, b–IR spectrum on the second day, c–IR spectrum on the third day).

The nanocomposite vibration emission in medium is determined by their dielectric characteristics and the corresponding functional groups presence in medium. At the metal/carbon nanocomposites obtaining the interaction of polymeric matrix with metal containing phase leads to formation of metal clusters covered by carbon shells accompanied by metal electron structure changes. In some cases the medium characteristics influence on nanocomposites throw into the increasing of nanocomposite surface energy portion, which concerns with changes of their electron structure and equally with electron structure of medium. In these cases the growth of metal atomic magnetic moment is observed that corresponds to the unpaired electron number increasing. The considerable changes of metal atomic magnetic moments in nanocomposites proceed when the phosphorus atoms include to carbon shells of nanocomposites. Thus, according to the analysis of metal/carbon nanocomposites characteristics, which are determined by their sizes and content, their activities are stipulated the correspondent dipole moments and vibration energies.

Metal/Carbon nanocomposite may be considered as "super molecule" which has the great dipole moment and vibrates in tera-Hz range. The vibration transfer within medium is ensured by the vibration propagation velocity (c) and depends on the medium density (ρ). At the same time the vibration propagation velocity within medium may be changed on the medium volume across which the vibration waves propagate on the distance l during τ time. Therefore, the energy of vibration (w) within medium is expressed.

$$W = \rho V\, 2\pi^2 v^2 \lambda^2 = m_{cp}\, 2\pi^2 v^2 \lambda^2 \qquad (12)$$

where V – volume, in which the nanocomposite vibration propagate, v and λ – the frequency and amplitude of nanocomposite vibration, m_{cp} – the average mass of medium in which nanocomposite vibration propagate, ρ – the medium density.

The metal/carbon nanocomposite has the great dipole and magnetic moments and vibrates in the field of medium molecules each from which has individual electric and magnetic moments. Therefore, according to [], metal/carbon nanocomposite is considered as vibrator, which radiates electromagnetic waves. For it the following equations, which describe mechanical and electromagnetic vibration are written –

$$D = A_m \sin 2\pi\, (t/P - l/\lambda) \qquad (13)$$

$$\Delta j = \varphi_m \sin 2\pi\, (t/P - l/\lambda) \qquad (14)$$

Parameters in equations are determined as: D – the chemical particles displacement to defined point within medium under the action of wave which is caused by metal/carbon nanocomposite; A_m – the amplitude of nanocomposite active vibration; t – the time of defined point achievement for vibration wave within medium; P – the period of vibration; l – the distance from "super molecule-metal/carbon nanocomposite" to defined point within medium in which the vibration wave pass;

l – the vibration wave length; Δj – the potential change in the medium point at the electromagnetic vibration propagation from "super molecule-metal/carbon nanocomposite"; φ_m – the maximum potential of electromagnetic radiation from metal/carbon nanocomposite.

The electromagnetic waves propagation velocity (W) depends on the medium properties.

$$W = \sqrt{[1/(\varepsilon m)]} \tag{15}$$

where e – dielectric constant, m – permeability of vacuum.

The electromagnetic wave propagation simultaneously with direct ultrasonic vibrations is accompanied by the orientation of charged chemical particles and dipoles within medium.

The electromagnetic waves are formed at the ultrasonic vibration of metal/carbon nanocomposites. These waves stimulate the changes of electron structure and the growth of magnetic moments of clusters within nanocomposites. It is shown that the nanocomposites vibration energies depend on their average masses. However, the specific surface of metal/carbon nanocomposites particles changes in dependence on the nature of nanocomposite in other order than the correspondent order of the vibration energies.

Therefore the energetic characteristics of metal/carbon nanocomposites are more important for the activity determination in comparison with their size characteristics.

28.4   CHANGES OF NANOCOMPOSITES PROPERTIES AT THE INTRODUCTION HETEROATOMS IN CARBON SHELLS OF METAL/CARBON NANOCOMPOSITES

28.4.1   INTERACTION OF NANOCOMPOSITES WITH AMMONIUM PHOSPHATES

There are several techniques to grate functional groups. By the bond strength the processes of connecting to nanostructures are divided into two groups: with the formation of strong covalent bonds and without such bonds (due to hydrophobic interaction, formation of hydrogen bonds). The method of obtaining phosphorus containing metal/carbon nanocomposites consists in the interaction of preliminary obtained metal/carbon nanocomposites with ammonium phosphates, in particular APPh, in specifically calculated proportion. On the experiment results, the optimal proportion for all three types of nanostructures is the correlation of ammonium phosphate (APh) or APPh to NC = 1:1. APPh interacts with NC in mechanical mortar following the experimentally found mode with water for better grinding. Afterwards the samples are dried in a dry heat oven and studied.

28.4.2 INVESTIGATIONS WITH X-RAY PHOTOELECTRON SPECTROSCOPY APPLICATION [7]

By the XPES data it was found that in some cases nanostructures before their modification with ammonium phosphates contained oxides of not completely reduced metals, the carbon structure was not completely formed (Fig. 5.5). After functional groups were grafted to nanocomposites (Fig. 5.6), a great number of bonds appropriate for nanostructures were formed, such as C-C, C-Me, the number of C-H bonds decreased. Below you can find the XPES spectrum and its analysis on the example of Fe/C nanocomposite.

Before the modification with APPh the process of iron reduction in nanocomposites is complicated and nanoproduct output is insignificant. The reduction process activity goes up after APPh is introduced. In this case, the metal is reduced and metal/carbon nanostructures are formed. After the modification C-Me bond is vivid on the spectra (Figs. 28.23 and 28.24).

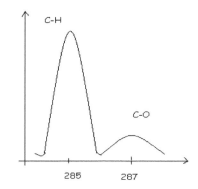

FIGURE 28.23   X-ray electron C1s spectrum of Fe/C nanocomposite before modification.

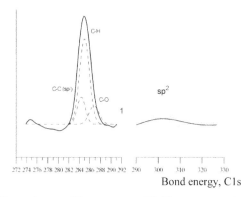

FIGURE 28.24   X-ray electron C1s spectrum of Fe/C nanocomposite after modification.

The sample magnetic moment increases indicating the structure improvement as, for example, the magnetic moment of iron/carbon NC increased from 2.2 up to 2.5 m_B, and nickel/carbon NC – from 1.8 up to 3 m_B; the magnetic moment of nickel reference sample was 0.6. Thus the modification results in the improvement of nanocomposite structures, metal reduction, appearance of additional functional groups increasing the nanostructure activity in materials.

28.4.3 INVESTIGATIONS BY IR SPECTROSCOPY [6]

In Fig. 28.25, the IR spectrum is seen on the example of copper/carbon nanocomposite. IR data demonstrate the appearance of the bands corresponding to phosphorus containing groups in the range 850–1250 cm^{-1}.

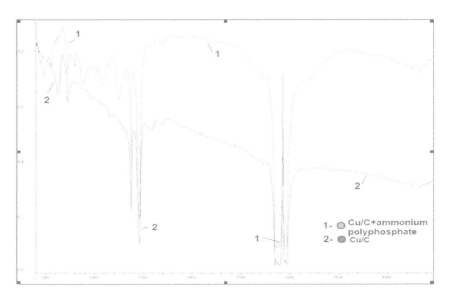

FIGURE 28.25 IR spectrum of modified copper/carbon nanocomposite (1) in comparison with non-modified nanocomposite (2)

IR spectrum indicates the peak of benzene ring flat deformation at wave number 1500–1600 cm^{-1}, π -electrons shift relative to the plane and benzene ring curves to the side of phosphoryl group contained in ammonium polyphosphate.

When comparing IR spectra of the samples, it should be pointed out that the absorption intensity on the spectrum of phosphoryl [18] sample is much greater (nearly in two times) in comparison with the sample spectrum, which does not contain phosphorus. This indicates the nanocomposite activity growth.

28.4.4 THE FUNCTIONALIZATION OF METAL/CARBON NANOCOMPOSITES BY NITROGEN CONTAINING COMPOUNDS [6]

The functionalization process is realized by means of mechanical and chemical activation during the joint grinding polyethylene polyamine with copper/carbon nanocomposite. IR investigation of Cu/C nanocomposite fine suspension confirms the quantum-chemical computational experiment regarding the availability of NC interactions with PEPA amine groups. The intensity of these groups increased in several times when Cu/C nanocomposite was introduced. IR spectra of PEPA and Cu/C NC FS were taken (Fig. 28.26).

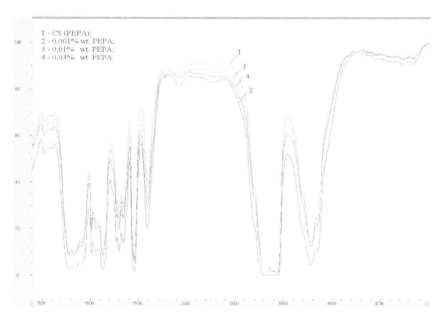

FIGURE 28.26   IR spectra of PEPA and FS of Cu/C nanocomposite.

The comparison of IR spectra of polyethylene polyamine and fine suspension of Cu/C NC on PEPA basis (Fig. 5.17) indicate that practically all changes of wave numbers in the spectra are within the error ± 2 cm^{-1}. However, in FS spectra the vivid increase in peak intensity corresponding to deformation oscillations of NH bonds is observed. These changes can spread onto the vast areas arranging a certain super-molecular structure, apparently involving the adjoining nitrogen containing groups into the process, which is demonstrated by the intensity change of these peaks.

28.5 INVESTIGATION OF METAL/CARBON NANOCOMPOSITES SUPER SMALL QUANTITIES INFLUENCE ON ACTIVE MEDIA

28.5.1 SELF ORGANIZATION OF ACTIVE MEDIA UNDER NANOCOMPOSITES ACTIONS

28.5.1.1 METHODS OF FINE DISPERSED SUSPENSIONS PREPARATION

To select the components of fine suspensions with the help of quantum-chemical modeling by the scheme described before [5], first, the interaction possibility of the material component being modified (or its solvent or surface-active substance) with metal/carbon nanocomposite is defined. The suspensions are prepared by the dispersion of the nanopowder in ultrasound station. The stability of fine suspension is controlled with the help of laser analyzer. The action on the corresponding regions participating in the formation of fine dispersed suspension or sol is determined with the help of IR spectroscopy. As an example, below you can see brief technique for obtaining fine suspension based on polyethylene polyamine.

IR spectra demonstrate the change in the intensity at the introduction of metal/carbon nanocomposite in comparison with the pure medium (IR spectra are given in Fig. 28.27). The intensities of IR absorption bands are directly connected with the polarization of chemical bonds at the change in their length, valence angles at the deformation oscillations, i.e. at the change in molecule normal coordinates.

When nanocomposites are introduced, the changes in the area and intensity of absorption bands, which indicates the coordination interactions and influence of nanostructure onto the medium are observed.

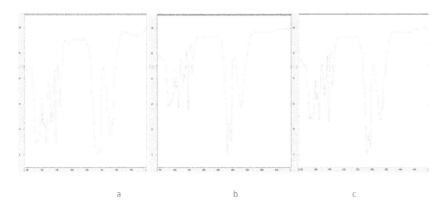

a   b   c

FIGURE 28.27 IR spectra of polyethylene polyamine (a), copper/carbon nanocomposite fine dispersed suspension (ω (NC) = 1%) (b), and nickel/carbon nanocomposite fine suspension (ω (NC) = 1%) (c) in polyethylene polyamine medium

Special attention in PEPA spectrum should be paid to the peak at 1598 cm^{-1} attributed to deformation oscillations of N-H bond, where hydrogen can participate in different coordination and exchange reactions. In the spectra wave numbers characteristic for symmetric $v_s(NH_2)$ 3352 cm^{-1} and asymmetric $v_{as}(NH_2)$ 3280 cm^{-1} oscillations of amine groups are present. There is a number of wave numbers attributed to symmetric $v_s(CH_2)$ 2933 cm^{-1} and asymmetric valence $v_{as}(CH_2)$ 2803 cm^{-1}, deformation wagging oscillations $v_D(CH_2)$ 1349 cm^{-1} of methylene groups, deformation oscillations of NH v_D (NH) 1596 cm^{-1} and NH$_2$ $v_D(NH_2)$ 1456 cm^{-1} amine groups. The oscillations of skeleton bonds at $v(CN)$ 1059-1273 cm^{-1} and $v(CC)$ 837 cm^{-1} are the most vivid. The analysis of intensities of IR spectra of PEPA and fine suspensions of metal/carbon nanocomposites based on it revealed a significant change in the intensities of amine groups of dispersion medium (for $v_s(NH_2)$ in 1.26 times, and for $v_{as}(NH_2)$ in approximately 50 times). Such demonstrations are presumably connected with the distribution of the influence of nanoparticle oscillations onto the medium with further structuring and stabilizing of the system. Under the influence of nanoparticle the medium changes which is confirmed by the results of IR spectroscopy (change in the intensity of absorption bands in IR region).

Density, dielectric penetration, and viscosity of the medium are the determining parameters for obtaining fine suspension with uniform distribution of particles in the volume. At the same time, the structuring rate and consequently the stabilization of the system directly depend on the distribution by particle sizes in suspension. At the wide range of particle distribution by sizes, the oscillation frequency of particles different by size can significantly differ, in this connection, the distortion in the influence transfer of nanoparticle system onto the medium is possible (change in the medium from the part of some particles can be balanced by the other). At the narrow range of nanoparticle distribution by sizes the system structuring and stabilization are possible. With further adjustment of the components such processes will positively influence the processes of structuring and self-organization of final composite system determining physical-mechanical characteristics of hardened or hard composite system. The effects of the influence of nanostructures at their interaction into liquid medium depend on the type of nanostructures, their content in the medium and medium nature

Similar changes in IR spectra take place in water suspensions of metal/carbon nanocomposites based on water solutions of surface-active nanocomposites. In Fig. 28.28, you can see IR spectrum of iron/carbon nanocomposite based on water solution of sodium lignosulfonate in comparison with IR spectrum of water solution of surface-active substance. As it is seen, when nanocomposite is introduced and undergoes ultrasound dispergation, the band intensity in the spectrum increases significantly. Also the shift of the bands in the regions 1100–1300 cm^{-1}, 2100–2200 cm^{-1} is observed, which can indicate the interaction between sodium lignosulfonate and nanocomposite. However, after two weeks the decrease in band intensity is seen. As the suspension stability evaluated by the optic density is 30 days, the nanocomposite activity is quite high in the period when IR spectra are taken. It can be expected that the effect of foam concrete modification with such suspension will be revealed if only 0.001% of nanocomposite is introduced.

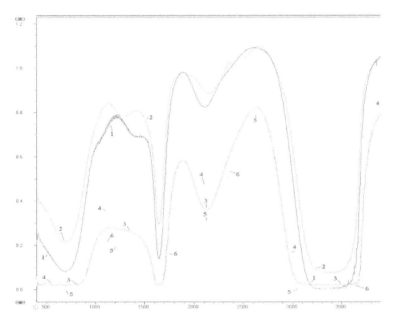

FIGURE 28.28   Comparison of IR spectra of water solution of sodium lignosulfonate (1) and fine suspension of iron/carbon nanocomposite (0.001%) based on this solution on the first day after nanocomposite introduction (2), on the third day (3), on the seventh day (4), 14th day (5) and 28th day (6).

The influence of nanostructures on the media and compositions was discussed based on quantum-chemical modeling [9]. After comparing the energies of interaction of fullerene derivatives with water clusters, it was found that the increase in the interactions in water medium under the nanostructure influence is achieved only with the participation of hydroxyfullerene in the interaction. The energy changes reflect the oscillatory process with periodic boosts and attenuations of interactions. The modeling results can identify that the transfer of nanostructure influence onto the molecules in water medium is possible with the proximity or concordance of oscillations of chemical bonds in nanostructure and medium. The process of nanostructure influence onto media has an oscillatory character and is connected with a definite orientation of particles in the medium in the same way as reagents orientate in nanoreactors of polymeric matrixes. To describe this process, it is advisable to introduce such critical parameters as critical content of nanoparticles, critical time and critical temperature [10]. The growth of the number of nanoparticles (n) usually leads to the increase in the number of interaction (N). Also such situation is possible when with the increase of n critical value, N value gets much greater than the number of active nanoparticles. If the temperature exceeds the critical value, this results in the distortion of self-organization processes in the composition being modified and decrease in nanostructure influence onto media.

28.5.2   POSSIBLE LIMITATIONS AT THE INTERACTION BETWEEN NANOCOMPOSITES AND MEDIA

The same limitations of metal/carbon nanocomposites applications for the media properties changes are possible. According to proposed mechanism of metal/carbon nanocomposite action on media and systems the limitations are determined by the properties of media and also the nature and characteristics of metal/carbon nanocomposites. It is noted the electromagnetic waves propagation velocity is decreased within liquid media having the great values of dielectric constant and permeability of vacuum. In these cases the probability of wave extinction is increased. The viscosity growth and also the growth of medium density influence on the propagation of mechanical elastic vibration from "super molecule – metal/carbon nanocomposite." At the same time the vibration energy depends on sizes and forms of nanoparticles as well as its mass. The increasing of vibration energy usually leads to the growth of electromagnetic radiation and electromagnetic waves propagation velocity. The transference of vibration energy across the nanocomposite electromagnetic waves on media is confirmed by the increasing of fine dispersed suspension IR spectra intensity. These suspensions contain super small quantities of metal/carbon nanocomposites. The stability of nanocomposite fine suspension changes in dependence of the medium nature, and the corresponding intensity of lines in IR spectra changes as well. The lines intensity increasing in IR spectra is conserved during the same time for different fine dispersed suspension and depends on their dielectric constant, permeability of vacuum and medium viscosity. After some time the intensity of these lines in IR spectra is decreased to level of lines intensity for liquid which does not contained the metal/carbon nanocomposite.

In addition the introduction of metal/carbon nanocomposites in liquid media, which are evaporated, leads to the activity decreasing of nanocomposite owing to its passivation because of vibration energy decreasing.

When the phase amount in composition changes at the hardening, for instance, because of the evaporation of active solvent, with which the metal/carbon nanocomposite interacts, the processes of coordination with other components take place. These processes can lead to the decreasing of nanocomposite influence on self-organization of systems and on the material properties. For example, the strength of concrete modified by the water fine suspension, containing 0,001% of Copper/Carbon nanocomposite, increases on 67% during 7 days. However, this difference decreases after 28 days to 43%.

The efficiency of metal/carbon nanocomposite can be low for modification of compositions in which the vibration energy of nanocomposite is less than the energy of medium activation.

In some cases this difference stipulates by the large density of medium, metal/carbon nanocomposites are very active in redox processes. Therefore, it is possible the changes in the structures of interacted media as well as in the structures of itself nanocomposites. These processes probably can lead to disturb self-organization of media under the nanocomposite action. In addition some nanocomposites change structures too because of the introduction the same atoms from media within the shell of nanocomposite.

The metal/carbon nanocomposite, which has the great dipole and magnetic moments, vibrates in the field of medium molecules each from which has individual electric and magnetic moments.

The competence of different nanostructures formation processes is possible. Among correspondent nanostructures can be one size, two size and three size super molecules (sub-molecules) in dependence on the conditions of interactions within medium or modification of compositions. Besides of possible different directions for these processes the mechanical characteristics do not increased in practice contrary thermal physical properties. Especially the great growth of thermal capacity takes place. At the same times the transference propagation of nanocomposite energetic influence on the medium molecules, accompanied with self organization of medium molecules, can be written with Avrami–Kolmogorov equation application. The problem of limitation for the nanostructures including the metal/carbon nanocomposites requires attention.

28.6   MODIFICATION OF POLYMERIC MATERIALS BY SUPER SMALL QUANTITIES OF METAL/CARBON NANOCOMPOSITES

28.6.1   THEORY OF MATERIALS MODIFICATION BY METAL/CARBON NANOCOMPOSITES

The essential changes of polymeric materials structures and properties at the metal/carbon nanocomposites addition in them are experimentally shown in papers. The hypothesis of corresponding nanostructures super small quantities influence on media through nanostructures vibrations transfer on media molecules when these vibrations near to ultrasound vibrations is proposed. Further this hypothesis is confirmed in some measure by the media IR spectra intensities increasing at the addition of metal/carbon nanocomposites in these media. In this case the self-organization of media molecules and the changes of corresponding media properties are found. In liquid media the self-organization effect depends on media viscosity and its polarity.

The action of super molecule nanocomposite electromagnetic radiation on polymeric polar compositions will be lead to the IR radiation increasing of correspondent media as well as to change of their electron state. It is noted that the nanocomposite super molecule surface energy vibration portion is only realized at its super small quantities addition into the correspondent polymeric systems. Besides the growth of lines intensity in IR spectra takes place. When there is the increasing of the nano-composite content in the polymeric systems, the growth of the rotation and electron portion of surface energy is observed. In this case the electromagnetic radiation part is decreased and the electromagnetic waves transference velocity is decreased too.

At the same time the relation of lines intensity (I) in IR spectral field (1300–1600 cm^{-1}, that correspond to nanocomposite skeleton vibrations) to the half width (a_h) of these lines characterizes the degree of polymeric systems order and also the nanocomposites activity in self-organization processes.

The working of IR spectra of fine dispersed suspensions and polymeric nano-structured films, modified by metal carbon nanocomposites, leads to the conclusion concerning to the activity growth and to the self-organization of polymer macromol-ecules at the nanocomposite quantities decreasing to 10^{-2}–10^{-40}%.

Below the IR spectra of polymethylmethacrylate films modified by super small quantities of Cu/C nanocomposite are represented (Fig. 28.29)

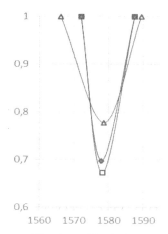

FIGURE 28.29   The changes of line intensities in IR spectra of polymethyl methacrylate films, which contains 10^{-3}% (□), 10^{-20}% (●), 10^{-10}% (▲) of copper/carbon nanocomposite

According Fig. 28.29, the decreasing of nanocomposite quantities leads to the increasing of line intensity at 1580 cm^{-1}. This growth corresponds to the vibration in-creasing of C=C groups under the action of nanocomposite electromagnetic waves.

Besides the correspondent relation –

$$D_2 = \frac{I_2}{a_{h2}}$$

(16)

for the system 2, modified by 10^{-3}% of metal/carbon nanocomposite, in comparison with analogous relation,

$$D_1 = \frac{I_1}{a_{h1}} \tag{17}$$

for the system, which contains 10^{-1}% of copper/carbon nanocomposite, is increased in more than two times, and the improvement of optical properties grows more than 20%.

Approximately the same level of changes is achieved on self-organization at the parameters comparison of polymeric systems and their modified analogues.

There is other way of estimation of self-organization degree, which is concluded in comparison of half width difference of lines in IR spectra.

In this case the decreasing of half width difference of lines in IR spectra is accompanied by the simultaneous growth of line intensity and corresponds to the increasing of self-organization degree.

For example, the self-organization of fine dispersed suspension (0.005% Cu/C nanocomposite based on polyethylene polyamine (2) in comparison with polyethylene polyamine (1) equals 38%, according to the calculation on formula.

The changes of media electro structure are possible under the nanocomposites super small quantities influence that it is confirmed by the X-ray photoelectron spectroscopic investigations.

Below C1s spectra for films of polyvinyl alcohol (PVA) and polyvinyl alcohol, modified by means of the addition of 0,001 and 0,0001% of Cu/C nanocomposite into liquid PVA, in comparison with the spectrum of very Cu/C nanocomposite (Fig. 28.30)

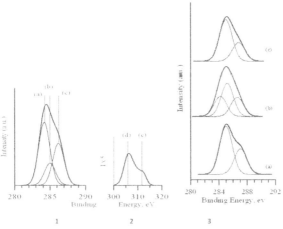

FIGURE 28.30 X-ray photoelectron C1s spectrum of Cu/C nanocomposite (1) and its satellites (2), and also the spectra for polyvinyl alcohol and its modified analogues (3).

1. C1s spectrum of copper/carbon nanocomposite: (a) for sp^2 hybridization (284 eV); (b) for carbon containing bonds (285 eV); (c) for sp^3 hybridization (286,2 eV).
2. The spectrum for satellites: (d) for sp^2 hybridization (306 eV); (e) for sp^3 hybridization (312 eV).
3. The spectrum for the film of polyvinyl alcohol (PVA) (a); the spectrum for film of PVA, modified by 10^{-3}% of Cu/C nanocomposite (b); the spectrum for film of PVA, containing 10^{-4}% of Cu/C nanocomposite (c).

The change of X-ray photoelectron spectrum of PVA, containing 10^{-3}% of Cu/C nanocomposite, is shown in the widening and in the appearance of satellites sp^2 and sp^3 lines at the simultaneous decreasing of C1s line (285 eV), which is known for C–H bond. It is possible that the reason of this fact is found in the increasing of polyvinylalcohol polarization under the nanocomposite action. It is possible to propose for the degree of self-organization of polyvinyl alcohol system the following formula:

$$DS = \frac{S_f - S_m}{S_f} \cdot 100\% \qquad (18)$$

where S_f – the fundamental peak area; S_m – the middle area of satellites.

For the considered case DS equals 55%, that corresponds to degrees of self-organization for the polymeric systems, modified by 10^{-3}% of Cu/C nanocomposite. This theoretical and experimental investigation will be continued for different models.

Thus, the methods of self-organization degree estimation in polymeric materials, modified by metal/carbon nanocomposites super small quantities, are proposed.

In addition the films of poly methyl met acrylate or polycarbonate, containing super small quantities of Copper/Carbon nanocomposite, are investigated. In them, beginning with the same concentration, the lines sp^2 and sp^3 hybridization from nanocomposite are observed. The levels of changes in x-ray photoelectron spectra (Fig. 28.31) depend on the Oxygen content in polymers and the polarization degree. This fact is confirmed by the increasing of refraction index for polycarbonate (PC) from 1.59 (non modified PC) to 1.65 (modified PC).

The results of IR and X-ray photoelectron spectroscopic investigations of poly methyl methacrylate and polyvinyl alcohol films modified by copper/carbon nanocomposite super small quantities are represented. The super small quantities change from 10^{-1} to 10^{-4}%. It is noted that the widening of C1s line in X-ray photoelectron spectra at the addition of 10^{-3}% of copper/carbon nanocomposite takes place. The increasing of lines intensity and decreasing of lines half width in IR spectra of analogous films of poly methyl methacrylate at the same concentration of Cu/C nanocomposite additives are observed. The increasing of lines intensity in IR spectra as well as the widening of C1s lines in X-ray photoelectron spectra at the addition of nanocomposite super small quantities in compositions is observed.

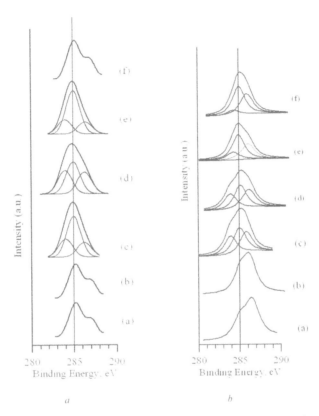

FIGURE 28.31 The XPS C1s-spectra of polymethylmethacrylate (*a*) and polycarbonate (*b*): (a) reference sample; (b) nanomodified with carbon copper-containing nanostructures in the amount of 10^{-1}%; (c) nanomodified with carbon copper-containing nanostructures in the amount of 10^{-2}%; (d) nanomodified with carbon copper-containing nanostructures in the amount of 10^{-3}%; (e) nanomodified with carbon copper-containing nanostructures in the amount of 10^{-4}%; (f) nanomodified with carbon copper-containing nanostructures in the amount of 10^{-5}%.

At the same time the intensity increasing in IR spectra of some media as well as the C1s lines widening in x ray photoelectron spectra is discovered. This fact can be explained by the media electronic structure changes, and also by the coordination interaction of nanocomposites with media molecules. The observed experimental results demand an answer with point of view concerning fundamentals of polymeric materials modification by super small quantities of metal/carbon nanocomposites. The metal/carbon nanocomposite vibration energy transference through electromagnetic waves on the medium molecules is confirmed by the increasing of intensity in IR spectra for the fine dispersed suspensions which contain the super small quantities of metal/carbon nanocomposite. The stability of correspondent suspensions changes in dependence on the medium nature. The essential changes of lines intensity in IR spectra take place for the polar liquids. The facts observed are correlated with growth of nanocomposites activities in the self organization processes at the materials production. The action of electromagnetic radiation of nanocomposite super molecule on polymeric polar compositions will be lead to the IR radiation increasing of correspondent media as well as to change of their electron state. It is noted that the surface energy vibration portion of nanocomposite super molecule is only realized at its super small quantities addition in the correspondent polymeric systems. Besides the growth of lines intensity in IR spectra takes place. When there is the increasing of the nanocomposite content in the polymeric systems, the growth of the rotation and electron portion of surface energy is observed. In this case the electromagnetic radiation part is decreased and the electromagnetic waves transference velocity is decreased too.

28.6.2 QUANTUM CHEMICAL MODELING OF MODIFICATION PROCESSES WITH METAL/CARBON NANOCOMPOSITES PARTICIPATING

28.6.2.1   QUANTUM–CHEMICAL INVESTIGATION OF INTERACTION BETWEEN FRAGMENTS OF METAL-CARBON NANOCOMPOSITES AND POLYMERIC MATERIALS FILLED BY METAL CONTAINING PHASE

Silver filler is more preferable due to excellent corrosion resistance and conductivity, but its high cost is serious disadvantage. Hence alternative conductive filler, notably nickel-carbon nanocomposite was chosen to further computational simulation. In developed adhesive/paste formulations metal containing phase distribution is determined by competitive coordination and cross-linking reactions.

There are two parallel technological paths that consist of preparing blend based on epoxy resin and preparing another one based on polyethylene polyamine (PEPA). Then prepared blends are mixed and cured with diluent gradual removing. Curing process can be tuned by complex diluent contains of acetiylacetone (AcAc), diacetonealcohol (DAcA) (in ratio AcAc:DAcA=1:1) and silver particles being treated with this complex diluent. Epoxy resin with silver powder mixing can lead to homogeneous formulation formation. Blend based on PEPA formation comprises AA and copper-/nickel-carbon nanocomposites introducing. Complex diluent removing tunes by temperature increasing to 150–160°C. Initial results formulation with

0.01% nanocomposite (by introduced metal total weight) curing formation showed that silver particles were self-organized and assembled into layer-chained structures.

Acetylacetone (AcAc) infrared spectra were calculated with HyperChem software. Calculations were carried out by the use of semiempirical methods PM3 and ZINDO/1 and ab initio method with 6-31G** basis set [36]. Results of calculation have been compared with instrumental measurement performed IR-Fourier spectrometer FSM-1201. Liquid adhesive components were preliminary stirred in ratio 1:1 and 1:2 for the purpose of interaction investigation. Stirring was carried out with magnetic mixer. Silver powder was preliminary crumbled up in agate mortar. Then acetylacetone was added. Obtained mixture was taken for sample after silver powder precipitation.

Sample was placed between two KBr glass plates with identical clamps. In the capacity of comparative sample were used empty glass plates. Every sample spectrum was performed for 5 times and the final one was calculated by striking an average. Procedure was carried out in transmission mode and then data were recomputed to obtain absorption spectra.

There are four possible states of AA: ketone, ketone-enol, enol A, enol B (Fig. 28.32). Vibrational analysis is carried out with one of semiempirical of ab initio methods and can allow to recognize different states of reagents by spectra comparison.

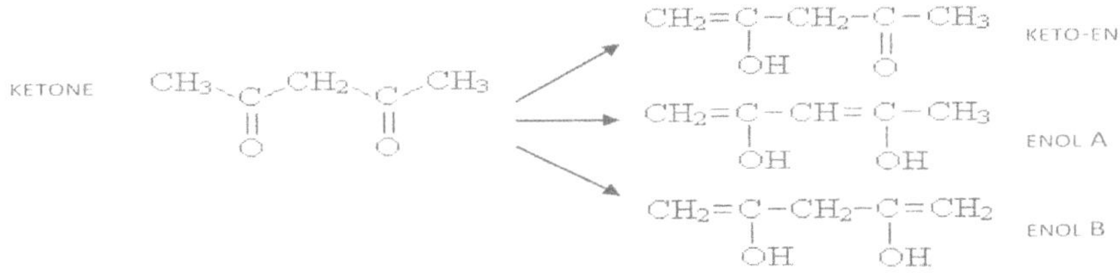

FIGURE 28.32   Different states of acetylacetone.

The vibrational frequencies are derived from the harmonic approximation, which assumes that the potential surface has a quadratic form. The association between transition energy ΔE and frequency v is performed by Einstein's formula

$$DE = h \times v \tag{19}$$

where h is the Planck constant. IR frequencies (-1012 Hz) accord with gaps between vibrational energy levels. Thus, each line in an IR spectrum represents an excitation of nuclei from one vibrational state to another [37]. Comparison between MNDO, AM1 and PM3 methods were performed in Ref. [38]. Semi-empirical method PM3 demonstrated the closest correspondence to experimental values. Figure 28.33 shows experimental data of IR-Fourier spectrometer AA measurement and data achieved from the National Institute of Advanced Industrial Science and Technology open database. Peaks comparison is clearly shown both spectra generally identical. IR spectra for comparison with calculated ones were received from open AIST database [39]. As it can be concluded from AIST data, experimental IR spectrum was measured for AA ketone state.

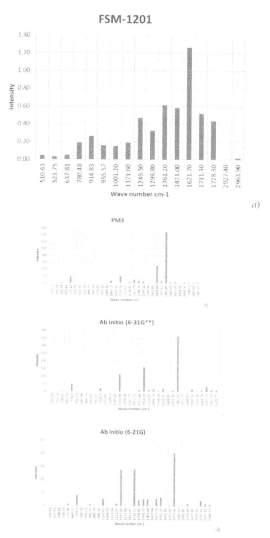

FIGURE 28.33 IR spectra of acetylacetone calculated with HyperChem software: semiempirical method ZINDO/1 (a); semiempirical method PM3 (b); ab initio method with 6-31G** basis set (c); ab initio method with 6-21G basis set (d).

Vibrational analysis performed by Hyper Chem was showed that PM3 demonstrates more close correspondence with experimental and ab initio calculated spectra than ZINDO/1. As expected, ab initio spectra were demonstrated closest result in general. As we can see at Fig. 28.33 (c, d), peaks of IR spectrum calculated with 6-21G small basis set are staying closer with respect to experimental values than ones calculated with large basis set 6-31G**. Thus it will be reasonable to use both PM3 and ab initio with 6-21G basis set methods in further calculations.

The optimal dimensions and shape of internal cavity of nanoreactors, optimal correlation between metal containing and polymeric components for obtaining the necessary nanoproducts are found with the help of quantum-chemical modeling. The availability of d metal in polymeric matrix results, in accordance with modeling results, in its regular distribution in the matrix and self-organization of the matrix, for example, silver cluster atom distribution in the fragment imitating the hardened epoxy resin (Fig.28.34).

FIGURE 28.34   IR spectra obtained with ab initio method with 6-21G basis set: acetylacetone (a); acetylacetone with silver powder (b); experimental data obtained from IR-Fourier FSM-1201 spectrometer: acetylacetone with silver powder in comparison with pure acetylacetone (c).

As shown on Fig. 28.34 (a, b), peaks are situated almost identical due to each other with the exception of peak 1573 cm-1 (Fig. 28.34, b) which was appeared after silver powder addition. Peak 1919 cm-1 (Fig. 7.15, b) has been more intensive as compared with similar one on Fig. 28.34 (a).

Peaks are found in range 600–700 cm^{-1} should usually relate to metal complexes with acetylacetone [40, 41]. IR spectra AA and AA with Ag+ ion calculations were carried out by ab initio method with 6-21G basis set. PM3 wasn't used due this method hasn't necessary parametrization for silver. In the case of IR spectra ab initio computation unavoidable calculating error is occurred hence all peaks have some displacement. Peak 549 cm^{-1} (Fig. 28.34, b) is equivalent to peak 638 cm-1 (Fig. 28.34, c) obtained experimentally and it is more intensive than similar one on Fig. 28.34. Peak 1974 cm^{-1} was displaced to mark 1919 cm^{-1} and became far intensive. It can be explained Ag+ influence and coordination bonds between metal and AA formation.

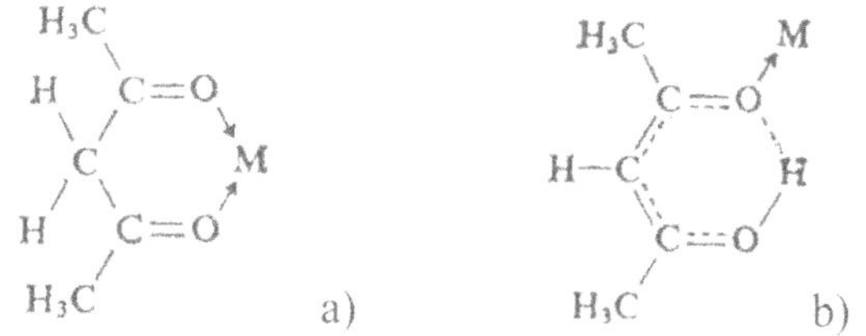

FIGURE 28.35   Different ways of complex between metal and AA formation: kentone with metal (a); enol with metal (b).

Complexes between silver and AA depending on enol or ketone state can form with two different ways (Fig. 28.35).

IR spectra of pure acetylacetone and acetylacetone with silver were calculated with different methods of computational chemistry. PM3 method was shown closest results with respect to experimental data among semiempirical methods. It was established that there is no strong dependence between size of ab initio method basis set and data accuracy. It was found that IR spectrum carried out with ab initio method with small basis set 6-21G has more accurate data than analogous one with large basis set 6-31G**. Experimental and calculated spectra were shown identical picture of spectral changes with except of unavoidable calculating error. Hence PM3 and ab initio methods can be used in metal-carbon complexes formations investigations.

The modification of materials by nanostructures including metal/carbon nanocomposites consists in the conducting of the following stages:

1. The choice of liquid phase for the making of finely dispersed suspensions intended for the definite material or composition.
2. The making of finely dispersed suspension with sufficient stability for the definite composition.
3. The development of conditions of the finely dispersed suspension introduction into composition.

At the choice of liquid phase for the making of finely dispersed suspension it should be taken the properties of nanostructures (nanocomposites) as well as liquid phase (polarity, dielectric penetration, viscosity).

It is perspective if the liquid phase completely enters into the structure of material formed during the composition hardening process.

When the correspondent solvent, on the base of which the suspension is obtained, is evaporated, the re-coordination of nanocomposite on other components takes place and the effectiveness of action of nanostructures on composition is decreased.

The stability of finely dispersed suspension is determined on the optical density. The time of the suspension optical density conservation defines the stability of suspension. The activity of suspension is found on the bands intensity changes by means of IR and Raman spectra. The intensity increasing testify to transfer of nanostructure surface energy vibration part on the molecules of medium or composition. The line spreading in spectra testify to the growth of electron action of nanocomposites with medium molecules. Last fact is confirmed by x-ray photoelectron investigations.

The changes of character of distribution on nanoparticles sizes take place depending on the nature of nanocomposites, dielectric penetration and polarity of liquid phase.

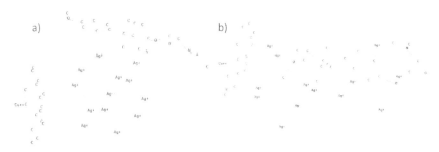

FIGURE 28.36 "Stretching" of silver nanocluster in the presence of nanostructure fragment and fragment of hardened epoxy resin molecule ED-20: (a) initial state; (b) result of system geometry calculation.

Form the given model it is seen that the nanocluster "stretches" along the fragment of hardened molecule ED-20 with changing the nanostructure position in such a way that the curved graphene plane consisting of 16 carbon atoms is placed between the oriented silver atoms and copper ion. The resulting picture can indicate the silver atom orientation in epoxy polymer fragment. At the same time, the metal orientation proceeds in interface regions and nanopores of polymeric phase which conditions further direction of the process to the formation of metal/carbon nanocomposite. In other words, the birth and growth of nanosize structures occur during the process in the same way as known from the macromolecule physics [21]

28.6.3 RESULTS OF MATERIALS MODIFICATION BY SUPER SMALL QUANTITIES OF METAL/CARBON NANOCOMPOSITES

28.6.3.1 GENERAL INFORMATION ABOUT THE MATERIALS PROPERTIES CHANGES AT THE MODIFICATION WITH METAL/CARBON NANOCOMPOSITES USING

For the production of materials with the improved characteristics the modification occurs with the using of Metal/Carbon Nanocomposites finely dispersed suspensions. The content of active nanocomposites usually makes up super small quantities (0.01–0.0001%). For even distribution of Metal/Carbon Nanocomposites super small quantities the number of suspensions are applied in correspondent compositions:

1. The water suspensions, containing nanocomposite and surfactant, are used for the modification of foam concrete and dense concrete.
2. The suspension based on the solution of PVC in acetone (or chlorinated paraffins) are used for the modification of PVC (polyvinyl chloride).
3. The suspensions based on the solution of PMMA in dichloroethane are used for the modification of PMMA (polymethyl methacrylate).
4. The suspension based on the solution of PC in methylene chloride (or dichloroethane) are used for the modification of polycarbonate (PC).
5. The suspensions based on the solution of phenol-formaldehyde resins in alcohol (or in mixture of alcohol and toluene) are used for the modification of glue BF-19.
6. The suspensions based on the polyethylene polyamine are used for modification of epoxy resins (cold hardening).
7. The suspensions based on the isomethyl tetrahydrophtalate anhydrate are used for modification of epoxy resins (hot hardening).

Before the application of the suspensions, they are usually diluted to the necessary quantity of nanocomposite. To select the components of fine suspensions with the help of quantum-chemical modeling by the scheme described before [1], first, the interaction possibility of the material component being modified (or its solvent or surfactant) with metal/carbon nanocomposite is defined. The suspensions are prepared dispersion of the nanopowder in ultrasound dispersator. The stability of fine suspension is controlled with the help of nephelometer (photocolorimeter) and laser analysator. The action on the corresponding regions participating in the formation of fine suspension or sol is determined with the help of IR spectroscopy. As an example, below you can see the brief technique for obtaining fine suspension based on polyethylene polyamine. IR spectra of metal/carbon and their fine suspensions in different (water and organic) media have been studied for the first time. It has been found that the introduction of super small quantities of prepared nanocomposites leads to the significant change in band intensity in IR spectra of the media. The attenuation of oscillations generated by the introduction of nanocomposites after the time interval specific for the pair "nanocomposite – medium" has been registered.

Thus to modify compositions with fine suspensions it is necessary for the latter to be active enough that should be controlled with IR spectroscopy.

Up to now we have carried out the laboratory and industrial experiments together with research and manufacturing divisions to modify inorganic and organic composite materials (concretes, foam concretes, epoxy binders, glue compositions, polyvinyl chloride, polycarbonate, polyvinyl acetate, fireproof coatings). A number of results of material modification with finely dispersed suspensions of metal/carbon nanocomposites are given, as well as the examples of changes in the properties of modified materials based on concrete compositions, epoxy and phenol resins, polyvinyl chloride, polycarbonate and current-conducting polymeric materials:

1. The introduction of metal/carbon nanostructures (0.005%) in the form of fine suspension into polyethylene polyamine or the mixture of amines into epoxy compositions allows increasing the thermal stability of the compositions by 75–100 degrees and also improving adhesive characteristics of glues and lacquers.

2. Hot vulcanization glue was modified by toluene-alcohol finely dispersed suspensions of copper/carbon or nickel/carbon nanostructures. On the test results of samples of four different schemes the tear strength σ_t increased up to 50% and shear strength τ_s – up to 80%, concentration of metal/carbon nanocomposite introduced was 0.0001–0.0003%.

3. The introduction of fine suspension of copper/carbon nanostructures (0.01%) leads to the significant decrease in temperature conductivity of the material (in 1.5 times). The increase in the transmission of visible light in the range 400–500 nm and decrease in the transmission in the range 560–760 nm were observed.

4. The PVC film modified containing 0.0008% of NC does not accumulate the electrostatic charge on its surface.

5. The introduction of nickel/carbon nanocomposite (0.01% of the mass of polymer filled on 65% of silver microparticles) into the epoxy polymer, hardened with polyethylene polyamine, leads to the decrease in electric resistance to 10^{-5} Ohm cm (10^{-4} Ohm cm without nanocomposite).

Thus, Metal/Carbon Nanocomposites, obtained in nanoreactors of polymeric matrixes, are effective for the different polymeric compositions.

Recently the unique properties of nanostructured composites formed in the process of nanoparticle introduction into composite materials have been attracting close attention of researchers. These properties are determined by nanometric sizes of the particles introduced and are not observed for bigger ones. Researches in this field demonstrate that nanostructured materials possess much better characteristics than similar compositions with chaotic particle layout. Nanoparticles are able to stimulate self-organization processes [1]. This feature of nanoparticles makes it possible to define the part of nanotechnology linked with their application as the technology that allows using the capability of nanoparticles to stimulate self-organization of systems for directed production of materials with the required properties.

It is rather perspective to use the nanocomposites obtained to improve the properties of composite materials. The introduction of such nanocomposites in super small quantities (0.0001–0.001% by mass) into materials has a positive effect on their structure and properties.

A more complex task during the modification of composite materials is the introduction of nanoadditives into the composite with uniform distribution of the additive in the material volume. Currently the obtaining of fine suspensions and colloid solutions of nanoparticles in various media is a widely spread and standard method for uniform distribution of nanoparticles.

Up to now we have carried out the laboratory and industrial experiments together with research and manufacturing divisions to modify inorganic and organic composite materials (concretes, foam concretes, epoxy binders, glue compositions, polyvinyl chloride, polycarbonate, polyvinyl acetate, fireproof coatings).

28.6.3.2 PROPERTIES OF EPOXY RESINS, GLUES AND PLASTICS BASED ON EPOXY RESINS, MODIFIED BY METAL/CARBON NANOCOMPOSITES

The compositions based on epoxy resins of cold and hot hardening for different purpose (binder of plastics, epoxy compounds, glues and coating) are modified by Metal/Carbon Nanocomposites. It is shown that the introduction of Cu/C nanocomposite (0.02%) into binder based on epoxy resins for the reinforced glass plastics production lead to their strength growth on 32.2% (Fig. 28.37).

FIGURE 28.37   The image of glass reinforced plastic armature.

The samples of epoxy polymer modified with Cu/C NC were produced mixing the epoxy resin heated up to 60°C and fine suspension of Cu/C NC on PEPA basis in proportion 10:1. The mixing took 5–10 min.

It is noted [16], that copper/carbon nanocomposite promotes the formation of well-regulated polymeric net with epoxy groups conversion near to 100% for the interval in two times less than analogous interval of hardening for the composition without nanocomposite.

The formation of well-regulated netted structure promotes to the substantial growth of heat capacity modified materials in comparison of epoxy material without nanocomposite (Fig. 28.38).

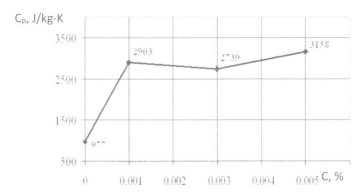

FIGURE 28.38   The modified epoxy resin heat capacity dependence on the copper/carbon nanocomposite quantities.

According to the results obtained of the heat capacity determination the heat capacity of epoxy material, modified by 0.005% Cu/C nanocomposite, on 223% more than heat capacity of non-modified analog.

The formation of coordinative bonds of nanocomposite with macromolecules of polymer, and also the formation of additional net, the increasing of degree of epoxy groups conversion and the formation of well regulated structure stipulates the increasing of destruction beginning temperature on 110° for epoxy polymeric material modified by 0.005% Copper/Carbon nanocomposite (Fig. 28.39).

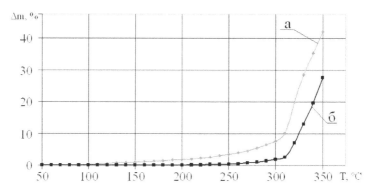

FIGURE 28.39 The epoxy polymers (non-modified – a, modified – b) mass losses dependence on temperature.

28.7   ADHESION

The adhesive strength of modified material in comparison with non modified analog to the such metals as copper and steel is studied. For the determination of adhesive strength standard method is used. The tests for defining the adhesion of modified epoxy resin to copper wire were carried out on tensile testing machine, the values of destruction load were found. The following diagram was prepared based on tests (Fig. 28.40)

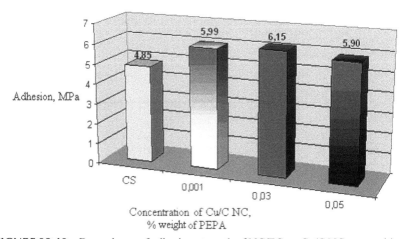

FIGURE 28.40   Dependence of adhesive strength of NC/EC on Cu/C NC composition.

The data from the diagram of adhesive strength (Fig. 28.14) indicate that the strength maximum of the modified epoxy resin is reached when Cu/C nanocomposite concentration is 0.03%. Probably the maximum strength of the modified epoxy resin at this concentration is conditioned by the optimal number of a new phase growth centers. The adhesion decrease with the concentration increase indicates that the number of Cu/C nanocomposite particles exceeded the critical value, which depends on their activity [10]. Therefore, probably the number of cross-links in the polymer grid increased – the material became brittle. The modification of industrial epoxy materials EZC–11 and ferric epoxy material with the using of 0.005% of Cu/C nanocomposite increases their adhesive strength more than 60%. These materials contain fillers, the particles of which are organized within well-regulated net formed under the influence of nanocomposite. In this case the formation of filler interacted particles chains is possible. Below the diagram of relative data on adhesive strength of named above industrial polymeric materials is given (Fig. 28.41).

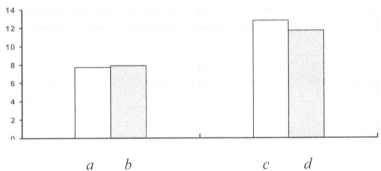

FIGURE 28.41 The relative characteristics of adhesive strength of epoxy material EZC-11(a) and ferric epoxy material (b) in comparison with their analogs (c, d) modified by 0.005% Cu/C nanocomposite.

28.7.1 THERMAL GRAVIMETRIC INVESTIGATIONS

For thermal gravimetric investigation also two sets of samples were produced, including the reference and modified FS of Cu/C NC with concentrations 0.001%, 0.01%, 0.03%, 0.05%.

28.7.1.1 THERMAL GRAVIMETRIC TECHNIQUE

Thermal stability was found by the destruction temperatures of modified epoxy resins. The temperatures of destruction beginning were determined by thermal gravimetric (TG) curves. Thermal balance DIAMOND TG/DTA was applied to obtain TG curves. TG curves of modified epoxy resins with NC concentrations 0.001%,

0.03%, 0.05% from PEPA weight were studied. The sample heating rate was 5°C/min.

To define the influence of nanocomposite on thermal stability of epoxy composition, a number of thermogravimetric investigations were carried out on reference and modified samples. The concentrations 0.001%, 0.03% and 0.05% from PEPA weight were used. Based on the results of thermogravimetric investigations the following diagram was prepared (Fig. 28.42).

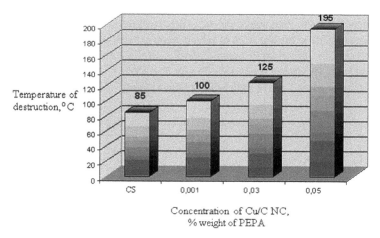

FIGURE 28.42   Dependence of thermal stability of cold-hardened epoxy composition on the concentration of copper/carbon nanocomposite.

When the concentration of nanocomposite elevates, the growth of polymer thermal stability is observed (Fig. 28.42). The thermal stability growth is apparently connected with the increase in the number of coordination bonds in epoxy polymer.

According to the investigation on the modification results of cold hardened epoxy resins the following conclusion may be made:

The test for defining the adhesive strength and thermal stability correlate with the data of quantum-chemical calculations and indicate the formation of a new phase facilitating the growth of cross-links number in polymer grid when the concentration of Cu/C nanocomposite goes up. The optimal concentration for elevating the modified epoxy resin (ER) adhesion equals 0.003% from ER weight. At this concentration the strength growth is 26.8%. At the same time the optimal quantity of Cu/C nanocomposite for elevating the modified industrial epoxy materials adhesion equals 0.005% that leads to the strength growth equals 60.7%. From the concentration range studied, the concentration 0.05% from ER weight is optimal to reach a high thermal stability. At this concentration the temperature of thermal destruction beginning increases up to 195°C.

The modification of hot hardened epoxy resins by mans of Metal/Carbon nano-composites is carried out with the application of the finely dispersed suspension based on isomethyl tetraphtalic anhydrate.

The modification of hot vulcanization glue with copper/carbon and nickel/carbon nanostructures to increase the level of adhesive characteristics was carried out with the help of fine suspensions produced based on toluene. After testing the samples of four different schemes, the increase in the strength at detachment σ_{det} up to 50% and shear τ_{sh} up to 80% was observed, the concentration of metal/carbon nanocomposite introduced was 0.0001–0.0003% (Fig. 28.43).

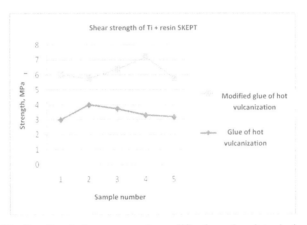

FIGURE 28.43 Results of glue compounds modification when introducing 0.0001% of copper/carbon nanocomposite.

The modification of conventional recipes of hot vulcanization glues with metal/carbon nanocomposites (51-К-45) leads to significant increase in adhesions charac-teristics on all glue boundaries investigated (Table 28.7).

TABLE 28.7 Results of Samples Tear and Shear Tests [9]

Tear strength σ_{tear}, kgs/cm^2			Shear strength τ_{shear}, MP a		
Conventional recipe of the glue 51-К-45	Modified 51-К-45		Conventional recipe of the glue 51-К-45	Modified 51-К-45	
	Ni/C	Cu/C		Ni/C	Cu/C
38.3	48.6	56.4	3.5	6.3	6.3

The application of these materials as adhesives for the gluing of metals and vul-canite is realized on the schemes "metal$_1$-adhesive$_1$-vulcanite-adhesive$_2$-metal$_2$." To define the adhesive tear and shear strengths the above proposed scheme was used (Figs. 28.44 and 28.45). The investigations carried out revealed that the modification of the conventional recipe of the glue 51-K-45 results not only in increasing the glue adhesive characteristics but also in changing the decomposition character from adhesive-cohesive to cohesive one.

The availability of metal compounds in nanocomposites can provide the final material with additional characteristics, such as magnetic susceptibility and electric conductivity.

The modification of different materials with super small quantities of metal/carbon nanocomposite allows improving their technical characteristics, decreasing material consumption and extending their application.

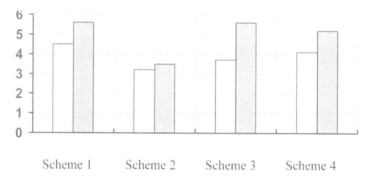

FIGURE 28.44   Relative changes of adhesive tear strengths of epoxy glues modified by Metal/Carbon Nanocomposites (content of NC – 0.0001%).

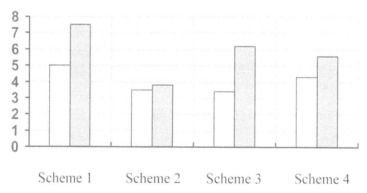

FIGURE 28.45   Relative changes of adhesive shear strengths of epoxy glues modified by Metal/Carbon Nanocomposites (content of NC – 0.0001%).

Thus, the application of Metal/Carbon Nanostructures is effective for the modification epoxy compounds, binders on the basis of epoxy resins for the reinforced glass plastics, different epoxy glues.

Results of modification of fireproof materials, fire resistant intumescent coatings and glues, modified by nanostructures. The modification of phenol-formaldehyde glues for the obtaining from them intumescent fire resistant glues with the using of Cu/C nanocomposites.

The glues based on phenol-formaldehyde resins (BF-19) were modified with copper/carbon nanocomposite and with phosphorylated analog. It was determined that nanocomposite introduction into the glue significantly decreases the material flammability. The samples with phospholyrated nanocomposites have better test results. When phosphorus-containing nanocomposite is introduced into the glue, foam coke is formed on the sample surface during the fire exposure. The coating flaking off after flame exposure was not observed as the coating preserved good adhesive properties even after the flammability test.

The nanocomposite surface phospholyration allows improving the nanocomposite structure, increases their activity in different liquid media thus increasing their influence on the material modified. The modification of coatings with nanocomposites obtained finally results in improving their fire-resistance and physical and chemical characteristics.

28.7.1.2 MODIFICATION OF THE GLUE BF-19 WITH FINE SUSPENSIONS OF METAL/CARBON NANOCOMPOSITES, INCLUDING PHOSPHORUS CONTAINING ONES

The glue BF-19 is intended for gluing metals, ceramics, glass, wood and fabric in hot condition, as well as for assembly gluing of cardboard, plastics, leather and fabrics in cold condition. The glue composition: organic solvent, synthetic resin (phenol-formaldehyde resins of new lacquer type), synthetic rubber.

When modifying the glue composition, at the first stage the mixture of alcohol suspension (ethyl alcohol + Cu/C NC_{mod}) and APPh was prepared. At the same time, the mixtures containing ethyl alcohol, Cu/C NC and APPh, ethyl alcohol and APPh were prepared. At the second stage the glue composition was modified by the introduction of phosphorus containing compositions prepared into the glue BF-19.

28.7.1.3 SAMPLES PREPARATION FOR FLAMMABILITY DETERMINATION

The samples to be tested are the plates with the dimensions $150 \times 15 \times 3$ (mm). The plates consist of foam polyethylene and paper glued together with phosphorus containing glue modified with metal/carbon nanocomposites with and without phosphorus. At the same time, check samples are prepared. These are the plates of foam polyethylene and paper glued together with the glue BF-19 filled with ammonium polyphosphate with phosphorus content in the glue 3, 4, and 5% from its mass.

28.7.1.4   TECHNIQUE OF SAMPLE FLAMMABILITY TESTING

When studying the influence of Cu/C nanocomposites on the flammability of polymeric coatings on the basis of phenol-formaldehyde resins (PFR) to select the optimal composition of nanostructures, the lengths of carbonized parts of the samples with 1.5-minute flame exposure were determined.

The composition of sample coating: glue BF-19 + APPh (Ammonium Polyphosphate).

To compare the results of coating flammability, three samples were selected and tested.

The test results revealed that the length of carbonized part of the samples containing APPh (Ammonium Polyphosphate) and exposed to burner flame for 1.5 minutes can be about 8.5 cm with 3% phosphorus content in the sample (Table 28.8).

TABLE 28.8   Results of testing samples containing Ammonium Polyphosphate

Sample No	Sample composition	Phosphorus content, %	Flame exposure time, min	Length of carbonized part of samples, mm
1	APPh	5	1.5	65.33
2	APPh	4	1.5	82
3	APPh	3	1.5	84.67
	Average			77.33

The tests of check samples confirmed that with the phosphorus content increase in composition the length of carbonized parts of samples goes down.

The composition of sample coating: glue BF-19 + APPh + Cu/C NC $_{pure}$.

The next step was to test samples containing nanocomposites without phosphorus content. The average value of the carbonized part of the samples was 21.81 mm. The test results (Table 28.9) allow making the conclusion that nanocomposite inclusion significantly decreases the material flammability (in 3.5 times).

TABLE 28.9   Results of Testing Samples of Modified Compositions

Sample No	Sample composition	Phosphorus content, %	NC content, %	Flame exposure time, min	Length of carbonized part of samples, mm
4	Cu/C NC $_{pure}$ + APPh	5	0.00025	1.5	15.33
5	Cu/C NC $_{pure}$ + APPh	4	0.0002	1.5	23.43
6	Cu/C NC $_{pure}$ + APPh	3	0.00015	1.5	26.66
	Average				21.81

The composition of sample coating: glue BF-19 + APPh + Cu/C NC mod.

Phosphorus containing samples of Cu/C NC had better flammability test results than samples with APPh and samples containing APPh and Cu/C NC$_{pure}$. The length of the carbonized part of the samples was less by 3 mm in the average [12]. The average value of the carbonized part of the samples was 18.89 mm (Table 28.10). Thus it can be concluded that the inclusion of pholsphorylated nanocomposite decreases the material flammability to a greater extent than non-pholsphorylated nanocomposite.

TABLE 28.10 Results of Testing Samples of Compositions Modified By Phosphorylated Nanocomposites

Sample No	Sample composition	Phosphorus content, %	NC content, %	Flame exposure time, min	Length of carbonized part of samples, mm
7	Cu/C NC $_{mod}$ + APPh	5	0.00025	1.5	14.67
8	Cu/C NC $_{mod}$ + APPh	4	0.0002	1.5	16.67
9	Cu/C NC $_{mod}$ + APPh	3	0.00015	1.5	25.33
	Average				18.89

From the data demonstrated based on the test results it can be concluded that nanocomposite inclusion into the glue composition significantly decreases the material flammability. The length of the carbonized part of the samples modified with nanocomposites was in 4.1 times in the average less in comparison with similar parameters of the samples not containing nanocomposites. The samples with phosphoryl groups in nanocomposites have better test results (Fig. 28.46).

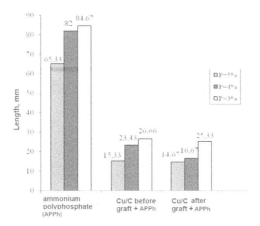

FIGURE 28.46 Diagram of the lengths of carbonized parts of the samples depending on phosphorus content in the composition.

The coating flaking off after flame exposure is not observed, that is, the coating preserves good adhesive properties even after the flammability test.

When the intumescent glue composition is modified with nanostructures, the material is structured with the formation of crystalline regions. In turn, such structuring under the influence of nanosystems results in the increased physical and mechanical characteristics, including their stability against high and low temperatures.

The application of metal/carbon nanocomposites is perspective for the modification of polymeric materials on a large scale as this is described in Refs. [9, 10, 12, 19].

28.7.2 THE PROPERTIES OF POLYVINYL CHLORIDE AND POLYCARBONATE, MODIFIED BY METAL/CARBON NANOCOMPOSITES ELECTRO CONDUCTED GLUES AND PASTS, MODIFIED BY METAL/CARBON NANOCOMPOSITES

The modification of polymeric films based on polycarbonate or polyvinyl chloride with the using of metal/carbon nanocomposites decreases the antistatic quantity essential for the substantial decrease of electrostatic charge on their surfaces. Especially, this is necessary for the PVC films.

The PVC film modified by the finely dispersed suspension based on chloroparaffins contains 0.0008% Fe/C nanocomposite and does not accumulate the electrostatic charge on the surface.

At the same time the crystalline phase in this material is increased. Below the image of final production of PVC films modified by Fe/C nanocomposite (0,0008%) is given in Fig. 28.47.

FIGURE 28.47   Final production stage of polyvinyl chloride film modified with Fe/C NC (0.0008%).

The material obtained completely satisfies the requirements applied to PVC films for stretch ceilings.

To modify polycarbonate-based compositions the Cu/C nanocomposite finely dispersed suspension based on mixture of methylene chloride and dichloroethane are produced. The introduction of 0.01% of copper/carbon nanostructures leads to the significant decrease in temperature conductivity of the material (in 1.5 times). The increase in the transmission of visible light in the range 400–500 nm and decrease in the transmission in the range 560–760 nm were observed.

The current conducted glues and pastes are obtained with the improved characteristics when the nickel/carbon nanocomposite is applied.

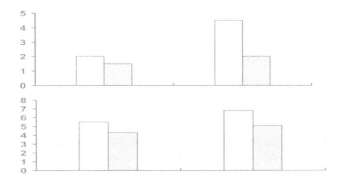

FIGURE 28.48   The tear and shear adhesive strength of current conducted epoxy material, modified by nickel/carbon nanocomposite.

TABLE 28.11   The Relative Results of Glues (Pastes) Current Conductivity Determination Modified By the Nickel/Carbon Nanocomposite

Parameters	Current conducted paste	Current conducted glue
Volume resistivity Ohm·cm (non modified)	2.4×10^{-4}	3.6×10^{-4}
Volume resistivity Ohm·cm (modified)	2.2×10^{-5}	3.3×10^{-5}

According to Fig. 28.48 and Table 28.11, the modification of polymeric materials by nickel/carbon nanocomposite leads to the increasing of their current conductivity as well as to the improving their adhesive properties. At the same time, the oscillatory nature of the influence of these nanocomposites on the compositions of foam concretes is seen in the fact that if the amount of nanocomposite is 0.0018% from the cement mass, the significant decrease in the strength of NC1 and NC2 is observed. The increase in foam concrete strength after the modification with iron/

carbon nanocomposite is a little smaller in comparison with the effects after the application of NC1 and NC2 as modifiers. The corresponding effects after the modification of cement, silicate, gypsum and concrete compositions with nanostructures is defined by the features of components and technologies applied. These features often explain the instability of the results after the modification of the foregoing compositions with nanostructures. Besides during the modification the changes in the activity of fine suspensions of nanostructures depending on the duration and storage conditions should be taken into account.

In this regard, it is advisable to use metal/carbon nanocomposites when modifying polymeric materials whose technology was checked on strictly controlled components.

At present a wide range of polymeric substances and materials: compounds, glues, binders for glass-, basalt and carbon plastics based on epoxy resins, phenol-rubber compositions, polyimide and polyimide compositions, materials on polycarbonate and polyvinyl chloride basis, as well as special materials, such as current-conducting glues and pastes, fireproof intumescent glues and coatings are being modified.

Below you can find the results of some working-out [9]:

1. The introduction of metal/carbon nanostructures (0.005%) in the form of fine suspension into polyethylene polyamine or the mixture of amines into epoxy compositions allows increasing the thermal stability of the compositions by 75–100 degrees and consequently increase the application range of the existing products. This modification contributes to the increase in adhesive and cohesive characteristics of glues, lacquers and binders.

2. Hot vulcanization glue was modified with copper/carbon and nickel/carbon nanostructures using toluol-based fine suspensions. On the test results of samples of four different schemes the tear strength σ_t increased up to 50% and shear strength τ_s – up to 80%, concentration of metal/carbon nanocomposite introduced was 0.0001–0.0003%.

3. Fine suspension of nanostructures was produced in polycarbonate and dichloroethane solutions to modify polycarbonate-based compositions. The introduction of 0.01% of copper/carbon nanostructures leads to the significant decrease in temperature conductivity of the material (in 1.5 times). The increase in the transmission of visible light in the range 400–500 nm and decrease in the transmission in the range 560–760 nm were observed.

4. Iron/carbon nanocomposite, the increase of the crystalline phase in the material was observed. The PVC film modified containing 0.0008% of NC does not accumulate the electrostatic charge on its surface. The material obtained completely satisfies the requirements applied to PVC films for stretch ceilings.

5. The introduction of nickel/carbon nanocomposite (0.01% of the mass of polymer filled on 65% of silver microparticles) into the epoxy polymer

hardened with polyethylene polyamine leads to the decrease in electric resistance to 10^{-5} Ohm cm (10^{-4} Ohm cm without nanocomposite).

In the paper the possibilities of developing new ideas about self-organization processes and about nanostructures and nanosystems are discussed on the example of metal/carbon nanocomposites. It is proposed to consider the obtaining of metal/carbon nanocomposites in nanoreactors of polymeric matrixes as self-organization process similar to the formation of ordered phases, which can be described with Avrami equation. The application of Avrami equations during the synthesis of nanofilm structures containing copper clusters has been tested. The influence of nanostructures on active media is given as the transfer of oscillation energy of the corresponding nanostructures onto the transfer of oscillation energy of the corresponding nanostructures onto the medium molecules.

IR spectra of metal/carbon and their fine suspensions in different (water and organic) media have been studied for the first time. It has been found that the introduction of super small quantities of prepared nanocomposites leads to the significant change in band intensity in IR spectra of the media. The attenuation of oscillations generated by the introduction of nanocomposites after the time interval specific for the pair "nanocomposite medium" has been registered.

Thus to modify compositions with fine suspensions it is necessary for the latter to be active enough that should be controlled with IR spectroscopy.

A number of results of material modification with fine suspensions of metal/carbon nanocomposites are given, as well as the examples of changes in the properties of modified materials based on concrete compositions, epoxy and phenol resins, polyvinyl chloride, polycarbonate and current-conducting polymeric materials.

KEYWORDS

- Functionalization
- Metal/Carbon nanocomposites
- Modification of systems and materials
- Quantum Chemical modeling
- Redox synthesis
- Self organization

REFERENCES

1. Kodolov, V. I., & Khokhrikov, N. V. (2009). Chemical Physics of the Processes of Formation and Transformation of Nanostrustures and Nanosystems. In two volumes *1*, 360, and *2*, 415, Izhevsk, Izhevsk State Agricultural Academy.

2. Shabanova, I. N., Kodolov, V. I., Terebova, N. S., & Trineeva, V. V. (2012). X Ray Electro Spectroscopy in Investigation of Metal/Carbon Nanosystems and Nano Structured Materials, *252*, Izhevsk-Moscow, Publ., "Udmurt University."

3. Kodolov, V. I., & Trineeva, V. V. (2013). Fundamental Definitions for Domain of Nanostrustures and Metal/Carbon Nanocomposites in Book, Nanostructure, Nanosystems and Nano-structured Materials, Theory, Production and Development, 1–42, Apple Academic Press, Toronto, Canada, New Jersey, USA.

4. Khokhriakov, N. V., Kodolov, V. I., Korablev, G. A. et al. (2013). Prognostic Investigations of Metal or Carbon Nanocomposites and Nanostructures Synthesis Processes Characterization, Ibid, 43–99.

5. Kodolov, V. I., Trineeva, V. V., Blagodatskikh, I. I. et al. (2013). The Nano structures Obtaining and the Synthesis of Metal or Carbon Nano Composites in Nanoreactors, Ibid, 101–146.

6. Kodolov, V. I., Akhmetshina, L. F., Chashkin, M. A. et al. (2013). The Functionalization of Metal or Carbon Nanocomposites or the Introduction of Functional Groups in Metal/Carbon Nanocomposites, Ibid, 147–175.

7. Shabanova, I. N., Terebova, N. S., Kodolov, V. I. et al. (2013). The Investigation of Metal or Carbon Nanocomposites Electron Structure by X-rays Photoelectron Spectroscopy, Ibid, 177–230.

8. Kodolov, V. I., Khokhriakov, N. V., Trineeva, V. V. et al. (2013). Computation Modeling of Nanocomposites Action on the Different Media and on the Composition Modification Processes by Metal/Carbon Nanocomposites, 231–286.

9. Kodolov, V. I., Lipanov, A. M., Trineeva, V. V. et al. (2013). The Changes of Properties of Materials Modified by Metal/Carbon Nanocomposites, Ibid, 327–373.

10. Kodolov, V. I., & Trineeva, V. V. (2013). Theory of Modification of Polymeric Materials by Super Small Quantities of Metal/Carbon Nano composites, Chemical Physics and Mesoscopy, *15(3)*, 351–363.

11. Kodolov, V. I., & Trineeva, V. V. (2012). Perspectives of Idea Development about Nanosystems Self Organization in Polymeric Matrixes in Book "The Problems of Nanochemistry for the Creation of New Materials," Torun, Poland, IEPMD, 75–100.

12. Akhmetshina, L. F., Lebedeva, G. A., & Kodolov, V. I. (2012). Phosphorus Containing Metal/Carbon Nanocomposites and their Application for the Modification of Intumescent Fire proof Coatings, Journal of Characterization and Development of Novel Materials, *4(4)*, 451–468.

13. Malinetsky, G. G. (2010). Designing of the Future and Modernization in Russia (Preprint of Keldysh, M. V. Institute of Applied Mechanics) *41*, 32.

14. Tretyakov, Yu. D. (2003). Self-Organization Processes (Uspekhi Chimii), *72(8)*, 731–764.

15. Kodolov, V. I., Kovyazina, O. A., Trineeva, V. V., Vasilchenko, Yu M., Vakhrushina, M. A., & Chmutin, I. A. (2010). On the Production of Metal/Carbon Nano Composites, Water and Organic Suspensions on their Basis, VII International Scientific-Technical Conference "Nanotechnologies to the Production, 2010," Proceedings Fryazino, 52–53.

16. Chashkin, M. A. (2012). Peculiarities of Modification by Metal/Carbon Nanocomposites for Cold Hardened Epoxy Compositions and the Investigation of Properties of Polymeric Compositions Obtained Thesis of Cand, Diss Perm, PNSPU, 17p.

17. Patent 2337062 Russia Technique of Obtaining Carbon Nanostructures from Organic Compounds and Metal Containing Substances, Kodolov, V. I., Kodolova, V. V. (Trineeva), Semakina, N. V., Yakovlev, G. I., Volkova, E. G. et al. Declared on 28.08.2006, Published on 27.10.2008.

18. Kodolov, V. I., Trineeva, V. V., Kovyazina, O. A. & Vasilchenko, Yu M. (2012). Production and Application of Metal/Carbon Nanocomposites, in Book "The Problems of Nanochemistry for the Creation of New Materials," Torun, Poland, IEPMD, 23–36.

19. Akhmetshina, L. F., Kodolov, V. I., Tereshkin, I. P., & Korotin, A. I. (2010). The Influence of Carbon Metal Containing Nano Structures on Strength Properties of Concrete Composites, Internet Journal "Nano Technologies in Construction," *6*, 35–46.
20. Kodolov, V. I., Khokhriakov, N. V., Trineeva, V. V., & Blagodatskikh, I. I. (2008). Activity of Nanostructures and its Display in Nanoreactors of Polymeric Matrixes and in Active Media, Chem. Phys. Mesoscopy, *10(4)*, 448–460.
21. Wunderlikh, B. (1979). Physics of Macromolecule in 3 Volumes, Mir., M, *2*, 574p.
22. Fedorov, V. B., Khakimova, D. K., Shipkov, N. N., & Avdeenko, M. A. (1974). To Thermo Dynamics of Carbons Materials, Doklady AS USSR, *219(3)*, 596–599.
23. Fedorov, V. B., Khakimova, D. K., Shorshorov, M. H. et al. (1975) To Kinetics of Graphitation, Doklady AS USSR, *222(2)*, 399–402.
24. Theory of Chemical Fiber Formation, Edited by Serkov, A. T. & Himiya, M. (1975) 548p.
25. Kodolov, V. I., Khokhriakov, N. V., Trineeva, V. V., & Blagodatskikh, I. I. (2010). Problems of Nanostructures Activity Estimation, Nanostructures Directed Production and Application, in Nanomterials Yearbook (2009). From Nanostructures, Nanomaterials and Nanotechnologies to Nanoindustry, N.Y. Nova Science Publ., Inc., 1–18.
26. Kodolov, V. I., Blagodatskikh, I. I., Lyakhovitch, A. M. et al (2007). Investigation of the Formation Processes of Metal Containing Carbon Nanostructures in Nanoreactors of Polyvinyl Alcohol at Early Stages, Chem. Phys. Mesoscopy, *9(4)*, 422–429.
27. Trineeva, V. V., Lyakhovitch, A. M., & Kodolov, V. I. (2009). Forecasting of the Formation Processes of Carbon Metal Containing Nanostructures using the Method of Atomic Force Microscopy, Nanotechnics, Iss., *4(20)*, 87–90.
28. Prigozhin, I., & Defay, R. (1966). Chemical Thermodynamics, Novosibirsk, Nauka, 510p.
29. Berezkin, V. I. (2000). Fullerenes as Embryos of Soot Particles, Physics of Solids, *42*, 567–572.
30. Khokhriakov, N. V., & Kodolov, V.I. (2005). Quantum-Chemical Modeling of Nanostructure Formation, Nanotechnics, *2, 108–112.*
31. Kodolov, V. I., Khokhriakov, N. V., Nikolaeva, O. A., & Volkov, V. L. (2001). Quantum-Chemical Investigation of Alcohols Dehydration and Dehydrogenization Possibility in Interface Layers of Vanadium Oxide Systems, Chem. Phys. Mesoscopy, *3(1), 53–65.*
32. Kodolov, V. I., Didik, A. A., Volkov, Yu A., & Volkova, E. G. (2004). Low-Temperature Synthesis of Copper Nanoparticles in Carbon Shell, Bulletin of HEIs, Chemistry and Chemical Engineering, *47(1)*, 27–30.
33. Lipanov, A. M., Kodolov, V. I., Khokhriakov, N. V. et al. (2005). Challenges in the Production of Nanoreactors for the Synthesis of Metallic Nanoparticles in Carbon Shells, Alternative Energetic and Ecology (ISJAEE), *2(22)*, 58–63.
34. Nikolaeva, O. A., Kodolov, V. I., Zakharova, G. S. et al. (20.03.2004). Methods of Obtaining Carbon-Metal-Containing Nanostructures, Patent of the RF 2225835.
35. Didik, A. A., Kodolov, V. I., Volkov, Yu A. et al. (2003). Inorganic Materials, *39(6)*, 693–697.
36. Famulari, A., Raimondi, M., Sironi, M., & Gianinetti, E. (1998). Chemical Physics, Iss., *232*, 289–298.
37. Spectral Database for Organic Compounds AIST. http://riodb01.ibase.aist.go.jp/sdbs/cgi-bin/cre_index.cgi?lang=eng.
38. Hyper Chem & manual (2002). Computational Chemistry, *149*, 369.
39. Schmidt, M. W., Baldridge, K. K., & Boatz, J. A. et al. (1993). Journal of Computational Chemistry, *14*, 1347–1363.
40. Seeger, D. M., Korzeniewski, C., & Kowalchyk, W. (1991). Journal Physical Chemistry, *95*, 68–71.
41. Kazicina, L. A., & Kupletskaya, N. B. (1971). Application of UV, IR and NMR Spectroscopy in Organic Chemistry, Moskow Vyshaya Shkola. 264.

INDEX

Milton Keynes UK
Ingram Content Group UK Ltd.
UKHW031141141024
449569UK00024B/1156

9 781774 632291